DINOSAURS
The Science behind the Stories

Judith G. Scotchmoor
Brent H. Breithaupt

Dale A. Springer
Anthony R. Fiorillo

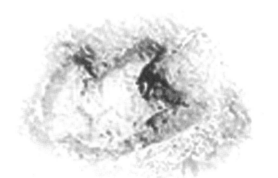

Society of Vertebrate Paleontology

The Paleontological Society

Published by American Geological Institute

Society of Vertebrate Paleontology

60 Revere Dr., Suite 500
Northbrook, IL 60062
(847) 480-9095
www.vertpaleo.org/

Founded in 1940, the Society of Vertebrate Paleontology has almost 2,000 members representing professionals, students, artists, preparators, and others interested in vertebrate paleontology. The Society is organized exclusively for educational and scientific purposes. The objective of the Society is to advance the science of vertebrate paleontology and to serve the common interests and facilitate the cooperation of all persons concerned with the history, evolution, comparative anatomy, and taxonomy of vertebrate animals, as well as field occurrence, collection, and study of fossil vertebrates and the stratigraphy of the beds in which they are found. The Society is also concerned with the conservation and preservation of fossil sites.

The Paleontological Society

Allen Marketing and Management
P.O. Box 1897
Lawrence, KS 66044-8897
(800) 627-0629 x215
www.paleosoc.org/

The Paleontological Society is an international organization devoted exclusively to the advancement of the science of paleontology through publications, meetings, funding opportunities, and outreach programs. The Society was founded in 1908 and today has about 1,700 members representing 40 countries. Members include professional paleontologists, academicians, explorationists, science editors, earth-science teachers, museum specialists, land managers, students, amateurs, and hobbyists. The Society subscribes to a Code of Fossil Collecting, which was overwhelmingly adopted by vote of membership in 1993.

American Geological Institute (Publisher and Distributor)

4220 King Street
Alexandria, VA 22302
(703) 379-2480
www.agiweb.org

The American Geological Institute (AGI) is a nonprofit federation of 40 scientific and professional associations that represent more than 120,000 geologists, geophysicists, and other Earth scientists. Founded in 1948, AGI provides information services to geoscientists, serves as a voice of shared interests in our profession, plays a major role in strengthening geoscience education, and strives to increase public awareness of the vital role the geosciences play in mankind's use of resources and interaction with the environment. The Institute also provides a public-outreach web site, www.earthscienceworld.org.

To purchase additional copies of this book or receive an AGI publications catalog, please contact AGI by mail or telephone, send an e-mail request to pubs@agiweb.org, or visit the online bookstore at www.agiweb.org/pubs.

Front Cover — Skeleton reconstruction, musculature, and life restoration of *Gryposaurus notabilis,* (R. Walters and T. Kissinger).

Design: De Atley Design
Layout: David K. Smith
Project Management: GeoWorks
Printing: CLB Printing Company

Contents

DEDICATION

We dedicate this volume to Edwin H. Colbert for his original research on

dinosaurs and for his devotion to sharing that research with the general

public. He inspired many of us who contributed to this book.

We further dedicate this book to all the K-12 teachers who share their

passion for science with their students.

Acknowledgments

Many persons have helped as we have assembled this book. We gratefully recognize David K. Smith for hours of detailed work on the layout and design, Julie De Atley for the cover design and the design concept of the book, and Julia A. Jackson for final edits and overall production. We thank the University of California Museum of Paleontology for donating both staff time and resources supporting production. In addition, we thank the Society of Vertebrate Paleontology and The Paleontological Society for supporting this effort and the American Geological Institute for its interest in publishing this volume. We also extend our sincerest thanks and appreciation to the following individuals for reviewing one or more of the chapters:

Steve Bailey	Sharon Janulaw
David Barrios	Sue Jagoda
Ericka Buehring	Jenny Lando
Kenneth Carpenter	Eric Lewis
Louise Cawthon	Cindy Maddox
Luis M. Chiappe	Anthony J. Martin
Daniel J. Chure	Brenda Massengill
Jennifer Collins	Jay Michalsky
Jim P. Diffily	Delinda Mock
Gigi Dornfest	Anne Monk
Jean Durrett	Homer Montgomery
Donna Engard	Brian K. Pilcher
Cecile Famosa	Sue Pritchard
David E. Fastovsky	Carmelo Sgarlato
Larry Flammer	Susan Sherman
Roland A. Gangloff	David K. Smith
Donald F. Glut	Mark Springer
Scott Hayes	Lara Sox-Harris
Sue Hoey	Glenn W. Storrs
Jack Horner	Mark Terry
David D. Gillette	Theresa DeLeon Weeks
Louis L. Jacobs	Colleen Whitney
Al Janulaw	Dale Winkler

And finally, we thank each of the authors who contributed to this volume. They all enthusiastically supported this publication and gave of their personal time to share their passion for dinosaur paleontology.

The Editors
Judith G. Scotchmoor, University of California Museum of Paleontology
Dale A. Springer, Bloomsburg University, Pennsylvania
Brent H. Breithaupt, University of Wyoming Geological Museum
Anthony R. Fiorillo, Dallas Museum of Natural History

Foreword

Dinosaurs and paleontology often provide the first steps that children take to learn more about the natural world. By the time those children begin school, more than likely they have mastered an "encyclopedic" knowledge of dinosaurs—their names, where they lived, when they lived, and on and on. They are primed for learning that *Tyrannosaurus* and *Velociraptor* are not beasts of mythology, but are part of a now extinct natural world, and that science provides a way of better understanding that world. They are primed for learning more about science.

This book is much more than another dinosaur book. It focuses on how we know what we know. It is meant for the teacher, the parent, and the grandparent who can help that eager child move beyond the visual portrayal of a dinosaur to the science that provides that image. By concentrating on science as a process, we can provide students with more than just the facts. We can teach them how science works and thus provide them with skills that will be useful for their lifetimes.

This book also satisfies the thirst of the "more mature" dinosaur buff. It arouses the imagination and inspires further questions. It reminds the reader that paleontology is more than dinosaurs. Paleontology provides a record of past environmental changes and life's responses to those changes. As such, paleontology can inform us about some of the dramatic patterns of change that are taking place today, about the disruptive influence of humanity on these patterns, and about how these changes may shape our future world.

Dinosaurs: The Science Behind the Stories is a collaborative effort of individuals who are passionate about their science and of three organizations which are equally passionate about providing the conduit through which that science can be shared — the Society of Vertebrate Paleontology, The Paleontological Society, and the American Geological Institute. As the presidents of each of these organizations, we would like to thank all who have contributed to this volume and we also thank the editors, Judy Scotchmoor, Dale Springer, Brent Breithaupt, and Tony Fiorillo, for their outstanding efforts in putting the book together.

We invite you to sit back, read, and enjoy.

Richard K. Stucky, *President, Society of Vertebrate Paleontology*
Patricia H. Kelley, *President, Paleontological Society*
Steven M. Stanley, *President, American Geological Institute*

Setting the Stage

I

1 Dinosaur paleontology provides an excellent opportunity to illustrate how science works. In this introductory chapter, **Springer and Scotchmoor** review the "rules" that separate science from non-science, but also stress that science is a dynamic and creative process. They give an overview of the focus of the book, which is to provide the evidence for how we know what we know about the history of life on Earth, particularly dinosaurs.

2 The nature of science and the place of historical sciences, such as paleontology and geology, are the focus of **Sampson**'s chapter. He explains the concept of "paradigm shifts" in science and shows how one of these major changes in perspective—plate tectonics—has accompanied a resurgence of interest in dinosaurs and their evolutionary history. As examples, he gives readers synopses of three areas of dinosaur research that have blossomed as a result of our new way of looking at these "terrible lizards."

PALEONTOLOGY

Unearthing the Past

Dale A. Springer
Department of Geography and Geosciences
Bloomsburg University, Bloomsburg, PA

Judith G. Scotchmoor
Museum of Paleontology
University of California, Berkeley

Walk into almost any elementary school classroom and ask about dinosaurs. The chances are pretty good that you will get an enthusiastic response. You are likely to be told that *Tyrannosaurus rex* was a ferocious meat-eater with nasty, sharp teeth and itty-bitty little arms; that dinosaurs laid eggs and took care of their babies after they hatched; and that a giant meteor killed off all the dinosaurs. Perhaps the students will tell you that birds are really dinosaurs. However, the obvious popularization of dinosaurs by the media also means that you might hear that *Pteranodon* was a flying dinosaur that carried off cavemen, that *Dilophosaurus* spat poison goo, or that dinosaurs were purple.

Certainly, if interest only were a measure of scientific knowledge, children would be Nobel Prize contenders when it comes to dinosaurs! But you might well raise an eyebrow at some of the things they tell you—things that may sound more like the result of too much television than the product of serious scientific inquiry. You would be correct to be skeptical. Some of the statements above have more "truth" to them than others. Better said: some are based upon actual evidence and logical inferences, and are thus more probably "correct" than the others. Some, on the other hand, are flat out wrong. But, which ones? How do we know what *T. rex* looked like, or how dinosaurs interacted with each other and their environment?

> *The question of "how do we know?" lies at the very core of the nature of science and is the focus of this book.*
>
> *Springer & Scotchmoor*

The question of "how do we know?" lies at the very core of the nature of science and is the focus of this book. In most sciences, we can propose hypotheses based upon our observations and then test them through experimentation. If many scientists perform the same tests and get the same results, we have established repeatability. Repeatability also means that we can now predict, with a high degree of confidence, the results of future runs of this particular experiment. The ability to make and test hypotheses and make predictions about the natural world is a fundamental aspect of science.

However, no scientist was alive when *Eoraptor* first appeared or when *Triceratops* roamed the plains of what is now North America. We cannot go back and "rerun" Earth history to see how dinosaurs evolved and diversified.

Dale A. Springer is a paleontologist and professor of Geosciences at Bloomsburg University in Pennsylvania. Her major research interests include the organization of marine invertebrate communities and investigating the processes responsible for changes in the spatial and temporal distributions of these communities. Dr. Springer has been involved in geoscience education initiatives for almost 25 years. She has served on several committees of the American Geological Institute (AGI), and is the current Education and Outreach Coordinator of the Paleontological Society (PS). She is an editor and co-author of a Paleontological Society publication, *Evolution: Investigating the Evidence*, and co-author of the booklet, *Evolution and the Fossil Record*, produced by

continued on p.5

So, how do paleontologists know anything about animals that have been extinct for 65 million years? Scientists must turn to different, but equally valid, ways of testing hypotheses and establishing repeatability.

The making and testing of hypotheses in historical sciences is based on the same fundamental premise as is every other science: the assumption that our senses—and the extensions of our senses, our scientific instruments—can give us accurate information about the natural universe. Every time we use our senses to examine the world around us, we are testing this premise and expanding our knowledge base. We are looking for patterns and trying to unravel the natural processes that created these patterns. Thus, we have learned that the physical universe operates according to certain physical and chemical laws, that certain aspects of the universe are constant within human experience. We can use our understanding of these natural laws, as well as data collected by scientists before us, to make hypotheses and inferences about the past. These ideas allow us to predict what we eventually expect to find in the fossil record. Such predictions are a way of testing the validity of our hypotheses, and they are every bit as useful and scientifically sound as any laboratory experiment.

Let's look at an example from dinosaur paleontology. Despite purported human footprints alongside those of dinosaurs and episodes of *The Simpsons* on television to suggest otherwise, the fossil record tells us that *Homo sapiens* and dinosaurs did not live at the same time. Observations made by thousands of scientists and amateur collectors over hundreds of years have never produced a bona fide fossil of *Homo sapiens* in the same rock formation as a fossil of a dinosaur. Each observation, each collection of fossils, represents an individual test of the prediction that we will not find dinosaurs and our species in rocks of the same age. We have achieved repeatability, established a pattern. We have a very high degree of confidence in our results. But we keep on testing, because the more times we repeat our test and obtain the same results, the more confidence we can place in the accuracy of our hypothesis: dinosaurs and humans did not co-exist.

© David K. Smith

The chapters of this book provide examples from dinosaur paleontology in which this same process is followed. Observations and tests from numerous lines of evidence (fossils, geology, chemistry, comparative anatomy, molecular biology) are used to explain, infer, test, and predict. Taken together, these steps have allowed us to attain a certain level of confidence about some aspects of the past. For example, we can see that birds are descended from theropod dinosaurs and that at least some dinosaurs show maternal behavior. In other cases, these steps are providing additional pieces to puzzles—such as thermoregulation in dinosaurs—that remain for us to unravel fully.

Science is all too often portrayed simply as a body of knowledge—a gazillion facts and figures collected over time by specialists, stored in neatly labeled boxes like butterflies in a museum drawer, just waiting to be pulled out, dusted off, and presented when needed. What an unenlightened, limited, and, well, boring way to view one of humankind's most exciting and wondrous activities! You might as well say art is merely a collection of paint, paper, stone, and wood, hanging in some hallway waiting to be viewed.

As illustrated within this book, science is far more than mere static information. Science is a process, a dynamic method of investigating the natural universe. Science is a work in progress, and it is self-correcting. It follows a very specific set of rules that define the boundaries that separate science from "not science." Yet, given these constraints, creativity, imagination, and even speculation still have a role to play, as long as we remain aware of their limits and employ them with due caution.

We must also remain aware that science is never done in an intellectual vacuum. Everyone, scientist or not, approaches a new experience or situation with preconceived ideas running around in their brains. This "mental baggage" is the cumulative result of an individual's life history—data, opinions, and cultural attitudes gained through direct experience or by learning from the experiences of others. Often, we do not fully realize or appreciate the extent to which our mental baggage influences our thoughts and actions. Most ordinary, everyday choices are made using this internalized information, and we would certainly be less able to make reasonable decisions without it. For example, it is much more painful to learn first-hand that fire burns than it is to learn and accept the information imparted in warning by your mother (although you could argue that there is something to be said for direct experience as a memory reinforcement).
On the other (unscorched!) hand, if we fail to recognize mental bag-gage, we risk overlooking poten-tial sources of bias in our thought processes. One of the hallmarks of good science is the ability to rec-ognize and to take into account such potential biases. Keep that in mind as you read this, or any other publication.

Few areas of science appeal to as wide an audience as dinosaur paleontology. Young or old, scien-tist or member of the general pub-lic, it is difficult to find anyone who has not been interested in di-nosaurs at some point during his or her life. Perhaps our fascination stems from our love of a good—but safe—fright. Dragons are fun, but fantasy; dinosaurs are real, but extinct. Neither has the remotest chance of truly causing us harm. Yet, we can, from the safety of our classroom or easy chair, travel to a world where "dragons" really ex-isted—in the bodies of dinosaurs. We can learn to see them not as menacing monsters or saccharine sweet Barneys, but as the com-plex, living, breathing, evolving animals they were.

© UCMP

continued from p. 3

the AGI and PS. She is also the President (2001–2002) of the Association for Women Geoscientists.

Judith G. Scotchmoor is Director of Education and Public Programs at the University of California Museum of Paleontology. Prior to this position, she was a 7th and 8th grade Science teacher for 25 years. With her teaching experience, it is not unusual that among her many roles at the museum, her primary interest is in the use of paleontology and technology as vehicles for improving science education in the classroom. She is currently the Project Coordinator of two National Science Foundation-funded programs—*Explorations Through Time* and *Understanding Evolution*. She serves on the Board of the Natural Science Collections Alliance, is Co-Chair of the Education Committee of the Society of Vertebrate Paleontology, and is editor and co-author of three resource books for teachers, *Learning from the Fossil Record, Evolution: Investigating the Evidence,* and *Making Connections*.

In the chapters of this book you will meet scientists and paleo artists who will share with you some of their passion, their enthusiasm, for dinosaurs. You will read about past and recent discoveries, follow paleontologists as they work through multiple lines of evidence that have led them to hypotheses about the anatomy, physiology, and behavior of dinosaurs, and learn how artists work with scientists to recreate Earth's past in light of current scientific knowledge.

Keep in mind as you explore these pages that dinosaur paleontology is a human endeavor, pursued by individuals with a keen desire to uncover the Earth's history. As such, each author brings to her or his article their own experiences, knowledge, and—possibly—their own agendas. Scientists communicate ideas through their publications; they expect colleagues to scrutinize their data and their methods. They want to convince other people of the validity of their hypotheses through the strength of the scientific arguments they present. Some can be very forceful in their efforts! There is nothing wrong with this, as long as you—their audience—read these articles with a mind of a skeptic, one who is willing to look at both the strengths and weakness of evidence presented and conclusions reached.

Finally, if you take away any messages from this book (besides the idea that dinosaurs are some of the coolest beasts that ever walked the Earth, and paleontologists have about the coolest job ever!), we hope they include these:

- Science is a dynamic process.
- Science belongs to everyone.
- Dinosaurs are a terrific way to introduce students of all ages to the nature and methods of science. And last, but far from least . . .
- Science is fun!

© UCMP

We would like to thank David K. Smith, UCMP, for enhancing this chapter with his artwork.

The Science of Paleontology: New Views on Ancient Bones

Scott D. Sampson
*Utah Museum of Natural History and Geology,
University of Utah, Salt Lake City*

2

As a dinosaur paleontologist, I am often confronted with a barrage of questions, such as "How do you know where to dig?" "Why do you think the dinosaurs went extinct?" Yet the commonly asked question that I find simultaneously most amazing and dismaying is, "Don't we already know everything about dinosaurs?" This question arises from a general lack of understanding about the nature of science. People typically think of science as a gradual accumulation of "facts" that has been ongoing for centuries. So it's commonly imagined that today, we scientists are simply adding a few grains of "factoids" to an enormous, firmly established mountain of knowledge. This view could hardly be further from the truth. Most of nature's secrets are still out there waiting to be discovered.

The overriding aim of science is to understand and describe as accurately as possible the workings of nature. In pursuit of this goal, certainty turns out to be a scarce commodity. Most scientists would agree that there is a single, physical reality we all seek to comprehend. To borrow the slogan of a recent popular television show, "the truth is out there." Yet the best we can offer are successive approximations of that truth.

The bottom line here is that scientific knowledge is not static, as many textbooks would have us believe, but highly dynamic and in constant flux. If in doubt, take the textbook challenge. Compare two texts from the same scientific topic, one current and the other about 25 years old. Regardless of scientific discipline, the books will almost certainly contain remarkable differences that show great leaps in our understanding. The misconception of fixedness or stability in knowledge is tragic in the sense that science is all the more exciting when we recognize that we don't have all (or even most) of the answers. Importantly, anyone, including children, can actually do science and make direct contributions to our understanding of nature.

Science provides a logical approach to addressing many kinds of problems, yet its reach is far from infinite. For example, science cannot assess the nature or existence of God. Nor, to take a trivial example, can it determine

> *. . . the communication of science should not revolve solely around what we think we know, but should also emphasize all that we don't know.*
>
> *Scott Sampson*

Scott D. Sampson is a Canadian paleontologist, who received his Ph.D. in Zoology from the University of Toronto in 1993. He is currently assistant professor in the Department of Geology and Geophysics and curator of vertebrate paleontology at the Utah Museum of Natural History, University of Utah. He has conducted paleontological fieldwork in a number of countries, including Zimbabwe, South Africa, Mexico, and Madagascar, as well as the United States and Canada. His research interests focus on the systematics, functional morphology, paleoecology, and biogeography of dinosaurs.

whether or not a television program, such as "Who Wants to be a Millionaire" is worthy of an Emmy Award. An integral part of science, as opposed to other ways of knowing the world, is hypothesis testing. Testability, and thus repeatability of results, is regarded as integral to the scientific method. Ideas must be framed as hypotheses that include testable predictions. Thus, science differs from other ways of knowing in that it relies on testability rather than belief. This means that everything in science is potentially open to scrutiny and modification.

In addition, science is undeniably a human pursuit, fraught with all the frailties of other such pursuits. Scientific interpretations can be heavily colored by one's own experience. Therefore, as in other arenas of life, two or more people can come to radically varying conclusions based on exactly the same evidence. Science, it must be remembered, is also inextricably embedded within culture, and therefore, it cannot help but reflect the current cultural milieu. A great proportion of work within the history of science is aimed at unveiling cultural biases that may have unconsciously dictated scientific inquiry and the nature of conclusions.

The Ladder of Confidence

In common parlance, the word "theory" is often assumed to imply speculation or conjecture. People then apply this connotation to science, with statements like, "Oh evolution, that's just a theory." Yet within scientific realms, to describe an explanation as a theory does not imply gross uncertainty. On the contrary, the theory label is used to refer to an idea that is robust, well tested, broad ranging, and supported by a large body of evidence.

Ideas within science fall into three major categories that can be viewed as a ladder of confidence. Imagine, however, that this ladder is somewhat oddly shaped; it consists of only three steps, the lowest being very narrow and the highest being extremely broad. Forming the lowest rung on the ladder are hypotheses, or formal expressions of ideas. Hypotheses are explanations of comparatively small scale,

sometimes preliminary in nature and sometimes strongly supported by a range of data. An example of a paleontological hypothesis is that birds are the direct descendants of dinosaurs. Once stated, various lines of evidence can be used to support or refute hypotheses.

Occupying the second rung on the ladder of confidence, a distant reach from the first, are scientific theories. A theory is a body of interconnected statements, based on extensive reasoning and testing, that account for a variety of observations. In contrast to hypotheses, theories tend to explain a relatively broad range of phenomena. An example is Einstein's theory of relativity, which, among other things, postulates that the speed of light is the upper limit for the velocity of any particle. This theory has been experimentally supported many times since, although aspects of it continue to be modified. Therefore, no theory is complete, and none are immune to change. That said, many theories, such as Darwin's theory of evolution, are so strongly supported that they approach being laws.

Finally, on the uppermost and most inclusive rung, representative of the highest level of confidence, are scientific aws. Laws are extremely rare. They are based on numerous observations of natural phenomena, and also explain a wide range of those phenomena. Examples are the law of gravity and the laws of thermodynamics. One of the key scientific laws used by paleontologists is the law of superposition, which states that in an undisturbed sequence of sedimentary rocks, the oldest rock layers will be at the base and the youngest will be on top.

Of course, any given idea cannot simply ascend this ladder of confidence ultimately to become a law. Rather, each successive level is occupied by ideas of increasing scope. Relative to hypotheses, theories are much more inclusive, with correspondingly broader implications. This means that hypotheses are the day-to-day currency of scientific interactions. It also means that true scientific theories are not poorly supported ideas or mere speculation. Rather, theories by definition reside on robust foundations of evidence. They are relatively broad, encompassing ideas that have withstood

the test of time, including many attempts to falsify them. Hypotheses, on the other hand, can be strongly or weakly supported by evidence, with some falling firmly in the realm of bald speculation.

It is always important to emphasize that a scientific hypothesis, theory or even law can never be fully proven, only supported or refuted by additional evidence. Ideally, investigators convey the level of confidence they have in a given hypothesis, separating those that can be regarded as probable, i.e., based on substantial supporting data, from those that are merely plausible, i.e., based on minimal data or simply a whim. Additionally, any scientist putting forth a particular hypothesis should address—and, preferably, rule out—as many alternative hypotheses as possible.

Paleontology as Science

Academic pursuits such as paleontology and geology are historical sciences, concerned predominantly with understanding and interpreting past events. Historical sciences stand in stark contrast to nonhistorical disciplines, such as physics and chemistry, in that scientists from the former camp are faced with a severe limitation; they typically cannot test a hypothesis through direct experimentation since it is impossible to reproduce past events. For example, barring the successful cloning of a dinosaur from its DNA (a highly unlikely event) or the invention of a time machine (even less likely), we clearly cannot investigate the metabolism of *Tyrannosaurus rex* directly. Similarly, geologists cannot observe the rifting and collisions of continents when endeavoring to reconstruct the pattern and timing of plate tectonic events. Reproducibility has been given high status within the ivory towers of science, and some have even argued that the inherent inability of historical sciences to reproduce results should dismiss them altogether from the realm of true scientific disciplines.

It turns out, however, that there is a way, and a quite elegant one at that, to circumvent, at least in part, the conundrum of time's arrow. Although the inexorable march of time prohibits actual reproduction of past events, it is pos-

sible to observe multiple examples of such events. If these examples are consistent with a given hypothesis, then that hypothesis gains support. If not, the hypothesis is falsified or at least is in need of modification. For example, every new specimen of primitive bird recovered in the fossil record serves as an independent test of the hypothesis that birds evolved directly from dinosaurs. In short, testability through multiple, independent examples is a hallmark of both paleontology and geology.

Paleontology is one of many hybrid disciplines within science, incorporating evidence and expertise from numerous other scientific disciplines, including zoology, botany, ecology, geology, chemistry, and physics, to name a few. It is in part for this reason that dinosaurs provide such an excellent vehicle for addressing many areas of science, and thus provide a tremendous teaching opportunity no matter what the age of the student. It is also for this reason that dinosaur paleontologists themselves come from disparate academic backgrounds. During most of the 20th century, the great majority of paleontologists were trained in university geology departments, arriving into the field with a good knowledge of rocks but little understanding of animals. More recently, there has been a strong trend toward training paleontologists in biology departments. Today, many paleontology students are completing graduate school with stronger backgrounds in such areas such as anatomy, ecology, and the study of historical relationships (known as phylogenetic systematics), as well as in geology.

A major problem faced by disciplines like dinosaur paleontology is the translation of science through the popular media. With notable exceptions—for example, certain science magazines and high quality television documentaries—the media rarely attempt to provide an in-depth portrayal of a given discovery. Bold or contentious claims often receive the lion's share of the limelight, with minimal discussion of the evidence, while the true complexities behind these claims fall victim to the editor's pen, or end up in pieces on the cutting room floor. For example, people were understandably astonished to learn that an asteroid single-handedly wiped out the dinosaurs 65 million

years ago. Here is a story with plenty of "bang for the buck," accounting not only for the untimely demise of dinosaurs and many other life forms, but also raising the specter of a repeat performance, with hapless humankind amongst the next potential victims. This idea is discussed briefly in a later section, but the main point here is that, even as a non-expert, it is important to be skeptical and ask pertinent questions. Yet how, you might be thinking, can I evaluate a scientific hypothesis unless I am at least reasonably familiar with the science? How do I know what questions to ask? Even without detailed knowledge of a particular topic, there are certain key queries to keep in mind. What assumptions have been made? What is the evidence supporting the claim? Have alternative hypotheses been explicitly considered and ruled out? In short, one must demand, "show me the evidence!"

Dinosaur Paleontology

Shaking the Foundation

Research within any field of science is conducted under a particular theoretical framework, often referred to as a "paradigm." Paradigms provide the conceptual infrastructure that guides scientific thinking within particular disciplines. Occasionally, a theory takes on grand proportions, causing an entire scientific field to reassess the very basis of its infrastructure. Such breakthroughs in science—for example, Darwin's theory of evolution by natural selection—invariably require, in the terminology of science historian Thomas Kuhn,[1] a "paradigm shift." Paradigm shifts entail a major restructuring, or even wholesale replacement, of an old theoretical framework.

The history of science is punctuated with numerous such intellectual shifts.[1] For the great Egyptian stargazer Ptolemy, the Earth was the center of the universe. Then an insightful Polish astronomer by the name of Nicholas Copernicus boldly placed the sun at the center of everything, forcing humans to regard their planetary home in this new light. Conversely, Sir Isaac Newton single-handedly devised a brilliant, mechanistic, clockwork-like perspective that formed the central paradigm of phys-

ics for hundreds of years. That is, until Albert Einstein revolutionized this paradigm by showing that Newtonian mechanics break down at speeds approaching that of light. The Einsteinian view turned out to have far greater predictive power under these special circumstances. Yet even Einstein's remarkable insights have not been immune to substantial revision.

So science within any discipline tends to be conducted within a prevailing paradigm, and it is often only following a paradigm shift that investigators begin to think about the world in wholly different ways, and devise innovative ideas to be tested. Think of science as a landscape, with its own distinctive topography of mountains, valleys and plains. Most of the time, this landscape is viewed as constant and relatively unchanging except for the passing of seasons. On very rare occasions, however, an earthquake shakes up the inhabitants (scientists) and forever alters the landscape. These paradigm shifts cause investigators to reassess basic assumptions and ask new kinds of questions.

An example of a recent paradigm shift, one that fundamentally impacted the study of dinosaurs and other ancient life, is the theory of plate tectonics, which was put forth in the 1920s but gained widespread acceptance only in the 1960s. This theory postulates that the Earth's surface is organized into large, mobile blocks or plates that ride atop a semi-molten layer beneath. These plates, including both continental and oceanic crust, diverge from one another along mid-ocean ridges, and collide violently along plate boundaries. Thus, the Earth is not a static body, but a dynamic sphere constantly undergoing upheaval. This realization accounted for numerous, disparate observations and completely revolutionized the study of geology. In the same way that evolution is recognized as the unifying theory of biology, plate tectonics quickly became the unifying theory of geology and geophysics.

Dinosaur remains have been recovered from every continent. How did these land-living creatures manage to disperse across oceanic barriers and populate the globe? Prior to plate tectonic theory, scientists devised all kinds of scenarios to account for the observed distribu-

tion patterns of dinosaurs and other animals. Today, of course, we realize that dinosaurs did not need to traverse oceans to spread around the globe. They simply hitched a ride on drifting continents (see Forster, page 45). When the dinosaurs first appeared during the Late Triassic, approximately 240 million years ago, all the continents were united into a single supercontinent known as Pangea. During the course of the Mesozoic Era—comprising the Triassic, Jurassic and Cretaceous Periods—Pangea fragmented into smaller landmasses. First there was Laurasia in the north and Gondwana in the south, and these blocks then broke apart, eventually resulting in the continents we recognize today. Life could not avoid being caught up in this super slow-motion dance of continental break-up and collision. The reason that dinosaurs were a global phenomenon is that they originated and diversified prior to and during the break-up of Pangea. As the continental rafts set sail on their various courses, the dinosaurian passengers on board, together with the rest of the flora and fauna, unknowingly went along for the ride. Once two blocks became separated, the plants and animals on each embarked on separate evolutionary journeys.

For example, the carnivorous dinosaurs (known as theropods) spread throughout Pangea during the Triassic and early Jurassic. Yet, later forms, those living in the Cretaceous, were restricted to particular geographic areas because opportunities for dispersal were limited by continental isolation. Thus, the Late Cretaceous tyrannosaurs, including *Tyrannosaurus* and its immediate kin, are found only in certain parts of the Northern Hemisphere. Meanwhile, down in the Southern Hemisphere, Late Cretaceous plant-eating dinosaurs faced different theropod predators with names like *Carnotaurus* and *Carcharodontosaurus*. Similarly, while we find closely related members of Jurassic-aged dinosaur groups—for example, the long-necked gargantuans known as sauropods (a.k.a. "brontosaurs")—as far apart as Wyoming and Tanzania, the diverse Late Cretaceous horned dinosaurs, such as *Triceratops*, are known only from western North America. So an understanding of plate tectonics and the movements of continents through

time is essential to understanding the geographic and evolutionary patterns we observe in dinosaurs.

Dinosaur Renaissance

Dinosaur paleontology experienced its own paradigm shift, one that also began in the 1960s. As a child of the baby-boom generation, my first exposure to paleontology occurred in the mid-1960s, when dinosaurs were generally regarded as sluggish, dim-witted behemoths. I fondly remember inspiring illustrations of sauropods fully submerged in lakes except for the tops of their snorkel-like heads. The logic was that these animals were simply too large to support their incredible masses on land, hence the aquatic lifestyle. Those dinosaurs that did walk on land were generally reconstructed as slow and awkward. Thus, the giant bipedal theropods like *Tyrannosaurus*, together with herbivorous forms such as the duckbills (hadrosaurs), were all reconstructed with upright bodies and massive tails dragging behind (Fig. 1A). Similarly, four-footed forms like the plant-eating *Stegosaurus* were portrayed with sprawled, lizard-like front limbs, low-slung bodies, and dragging tails (Fig. 1C). The overall impression was one of gigantic creatures lumbering across the landscape, with brains barely sufficient to carry them from day to day.

Then, within just a few years, dinosaurs were literally reinvented. Sauropods were no longer pictured submerged in water, but instead walking on land with columnar, elephantine-style limbs (actually, an old idea that once again came into vogue). Indeed, scientists were later to argue that an aquatic lifestyle would have been impossible for these long-necked giants, since water pressure at the depth of the chest would have compressed the chest cavity and severely restricted breathing. Meanwhile, *T. rex* and the other theropods, together with bipedal herbivores like the hadrosaurs, were reconstructed with entirely new postures, possessing horizontal rather than upright bodies (Fig. 1B). The tail, no longer trailing uselessly behind, was now held aloft, projecting rearward to act as a counterbalance for the head and trunk. This new look is strongly suggestive of much more active, agile animals. *Stegosaurus* and its four-footed kin were also transformed, bestowed with upright limbs and an airborne, nimble, potentially lethal tail (Fig. 1D). In addition to highly modified bodies, the "post-shift" dinosaurs were regarded as considerably more intelligent, with associated complex behaviors. Thus emerged hypotheses of pack hunting in some carnivores, together with herbivores exhibiting parental care and herding. The overall effect was to make dinosaurs more

Fig. 1. Changing views of dinosaurs before and after the paradigm shift in dinosaur paleontology. **A.** *Tyrannosaurus* prior to the paradigm shift (after ca. 1950 painting by Rudolf Zalinger, in Barnett;[23] **B.** *Tyrannosaurus* after the paradigm shift (after Paul[24]); **C.** *Stegosaurus* before the paradigm shift (after 1901 painting by Charles Knight, in Czerkas[25]); **D.** *Stegosaurus* following the paradigm shift (after Czerkas[25]).

akin to warm-blooded birds ad mammals than to cold-blooded reptiles.

What happened to bring about this fundamental change in our conception of dinosaurs? It was a paradigm shift, one triggered by a combination of discovery and insight. The fossil discovery was none other than the original sickle-clawed "raptor" theropod, recovered in Montana in 1964 by an expedition from Yale University. The revolutionary insights came from Yale paleontologist John Ostrom. In his 1969 description of this extraordinary carnivore, which he called *Deinonychus* ("terrible claw"), Ostrom argued that at least some dino-

saurs were considerably more active than previously assumed.[2] Shortly thereafter, noting a large number of bird-like features on the skeleton of *Deinonychus* and other theropod dinosaurs,[3,4] he reawakened the idea that birds evolved from dinosaurs and thus were, in a very real sense, dinosaurs themselves (see Currie, page 89).

Forging the bird-dinosaur link has entailed extensive study and detailed anatomical comparisons. Ostrom catalogued numerous characteristics linking theropod dinosaurs with birds. Subsequent workers have added many more, bringing the total number of shared, derived features between dinosaurs and birds to greater than 100.[5] Most experts today would argue that, beyond any reasonable doubt, birds are in fact theropod dinosaurs. Indeed, it appears that many features traditionally associated with birds evolved in the pre-bird, theropod ancestor. For example, like birds, theropod dinosaurs have thin-walled limb bones and a variety of air-filled (pneumatic) bones in the skeleton. Even feathers, the supposedly quintessential avian feature, have recently been found preserved on a variety of nonavian theropods from China, including the "raptor-like" forms.[6] Importantly, establishment of the intimate historical link between dinosaurs and birds has answered an age-old question. What did dinosaurs taste like? Like chicken, of course!

Faced with this new evidence and a fresh perspective, paleontologists quickly began to view dinosaurs as more bird-like than lizard-like. Scientists returned to the same evidence available previously and reassessed long-held views and biases. Soon the skeletons of bipedal dinosaurs like theropods and hadrosaurs were re-designed to assume a horizontal rather than an upright posture. Support for the new posture, and those for other dinosaurs, was discerned on the basis of various bony features, but also from footprints and trackways, most of which exhibited no evidence of a dragging tail. In addition, as with any paradigm shift worth its salt, entirely new questions were asked, in turn spawning novel research programs and heated debates. Were dinosaurs warm-blooded? Did some dinosaurs exhibit parental care? What were the intellectual and

behavioral capacities of the different dinosaur groups? In an attempt to address these questions, paleontologists have applied a range of analytical tools old and new, from detailed anatomical comparisons with living animals to Computer Tomography (CT) scanning. The end result has been several decades of extremely active research, accompanied by numerous key insights, many of which are discussed in this volume.

Nonetheless, it is often true of cultural trends that the pendulum, once in motion in a particular direction, tends to swing to a great extreme; this has certainly been the case with dinosaurs. If Ostrom's work was the spark that ignited the paradigm shift, then the fuel for the subsequent explosion of work came from Robert Bakker, a flamboyant ex-student of Ostrom, who championed the dinosaur renaissance.[7] Not long after Ostrom's original argument for more active, potentially warm-blooded dinosaurs, Bakker began reconstructing sauropods rearing up on their hind legs in order to battle marauding theropods. Similarly, *Tyrannosaurus* and its large carnivorous kin, no longer awkward and lumbering behemoths, were depicted as agile predatory machines capable of running speeds in excess of 40 miles per hour. Then there were the small raptor-like, sickle-clawed theropods like *Deinonychus* and *Velociraptor*, traveling in packs and utilizing a combination of cunning and cooperative behavior to take down prey of much greater body sizes. The *Jurassic Park* movie series brought these new ideas to popular audiences via the big screen, of course stretching the science to even greater extremes. Indeed, dinosaur science and dinosaur media have sometimes become blurred in recent years.

Today, we are beginning to see a reverse swing of the pendulum, as paleontologists use the same limited dataset to generate more tempered reconstructions of dinosaurian lifestyles. One example is the recent work indicating that tyrannosaurs and other large theropods could not attain the remarkable, jeep-pursuing speeds previously reported, and that they may have been incapable of true running. However, in contrast to large mammals in modern ecosystems, large theropods likely had no need for

such locomotor prowess, since they were still able to achieve substantially greater speeds than their herbivorous prey.[8]

Exploring the Changes in Dinosaur Research

Here are three examples of research areas that have exploded in the wake of the dinosaur renaissance. All of these, along with many others not mentioned here, are described in more detail in other contributions within this volume (see de Ricqlès, page 79; Dodson, page 153; Archibald, page 99).

Hot-Bloods or Cold-Bloods?

The supercharging of dinosaurs ignited a heated debate on dinosaur metabolisms. Ostrom argued forcefully that the high activity levels indicated by the bones suggested the possibility of warm-blooded dinosaurs. By this he meant that, like living mammals and birds, dinosaurs might have been "endothermic," maintaining relatively constant body temperatures despite varying environmental conditions. This thermal strategy is contrasted with cold-blooded, or "ectothermic," animals (most fish, amphibians, and reptiles), which have lower metabolic and body core temperatures that vary with the ambient temperature.

The chemical processes of metabolism are speeded up at higher internal body temperatures, so increased metabolic rates mean greater levels of activity. The benefits of endothermy include a greater potential for constant activity—useful for catching prey and/or escaping capture—and the ability to be fully active at night. Another advantage of warm-bloodedness relates to activity level. Due to differential rates of oxygen consumption, endotherms are capable of much higher levels of activity sustained over longer periods than are ectotherms. For example, humans or antelope can run continuously for great distances whereas lizards and crocodiles are restricted to shorter bursts of activity.

We generally think of endothermy as the best condition, undoubtedly because we ourselves (like nearly all other mammals) maintain constant body temperatures. However, the

benefits of warm-bloodedness come only at great cost. In order to keep the "high octane" internal furnace stoked, endotherms require about 10 times the amount of food of a similar sized ectotherm. For example, a meat-eating mammal like a lion must eat its own body weight in prey every nine days while a similar sized carnivorous reptile like the Komodo dragon has only to consume the equivalent of its body weight every 90 days. So endotherms must eat often and in great quantities.

What about dinosaurs? The closest living relatives of dinosaurs are crocodiles and birds. On the one hand, crocodiles are good "cold-blooded" ectotherms. They must lie in the sun to warm up before they can be active, and their potential for activity levels even at midday is much less than a similar sized mammal. Birds, on the other hand, are warm-blooded, or endothermic, allowing them to be active day and night, and to be active for longer periods than similar-sized reptiles.

Lacking any opportunities to measure dinosaur body temperatures or metabolisms directly, we must rely on circumstantial evidence, which is rarely conclusive. Numerous lines of evidence have been tapped in the search for the "Rosetta stone" of dinosaur metabolism.[9] These lines include posture and gait (more like mammals and birds than reptiles), relative brain size (approaching bird and mammal proportions in some dinosaurs, especially the small theropods), predator/prey ratios (thought by some to resemble most closely ratios in living mammals), and the existence of dinosaurs at high latitudes (suggestive of the ability to exist in colder temperatures).

One of the most interesting and controversial lines of evidence has been the microstructure of bone.[10] For example, the bones of juvenile dinosaurs indicate that at least some species grew very fast. Rapid growth rates suggest a rapid metabolism and warm-bloodedness. Mammals and birds also tend to possess abundant channels for blood vessels in the outer layers of their bones. This pattern, known as "Haversian bone," is related to the greater blood supplies necessary to maintain an endothermic metabolism. In contrast, the outer layers of reptile bones are often charac-

terized not by Haversian bone but by lines indicative of periodic slowed growth akin to tree-rings. It turns out that the situation is much more complicated than originally hoped. Haversian bone is sometimes present in large ectotherms such as crocodiles, whereas growth lines have been observed in a range of birds and mammals.

In the quest to unravel metabolic strategies of dinosaurs, body size is a key concept. Large animals have relatively greater volume and less surface area than smaller animals, which in turn causes the larger forms to retain heat longer. Galapagos tortoises, for example, weigh in at about 450 lbs. and can maintain a body temperature several degrees higher than ambient even on chilly nights. This strategy, in which body temperature is maintained by virtue of size rather than greater metabolic rates, is sometimes referred to as "gigantothermy." Control of body temperature via gigantothermy may well have characterized some of the large dinosaurs.

All in all, the debate over warm-bloodedness in dinosaurs has thus far resulted in considerably more heat than light. However, the research has taught us much about the metabolic strategies of all animals, not just dinosaurs. At this point, it seems likely that dinosaurs, like mammals, possessed a range of strategies for controlling metabolism and activity. Some may have been endothermic, others ectothermic, and still others may have had intermediate physiologies with intermediate metabolic rates.[11] Determining the exact nature of those strategies will keep paleontologists busy for some time to come.

Bizarre Structures

A large number of dinosaurs possess bizarre bony structures in their skeletons. For example, there are the horns and frills of horned dinosaurs (ceratopsians), the crests of duck-billed dinosaurs (hadrosaurs), the spikes and plates of stegosaurs, and tail-clubs of ankylosaurs. Various ideas have been put forth to explain these features. Traditionally—that is, prior to the paradigm shift—it was generally thought that they were used as weapons in defense against

theropod predators. Yet many of the theropods also exhibit their own bizarre structures, typically taking the form of horns or crests protruding from the top of the skull. A more recent explanation is that at least some of these bony appendages functioned in control of body temperature, or thermoregulation.[12,13] An alternative is that many of these elaborate structures functioned first and foremost in courtship and combat with members of the same species; that is, they were primarily signals used to identify and compete for mates. This idea was first articulated just prior to the dinosaur renaissance,[14] but has since received broad support.[15,16,17]

Since we cannot observe dinosaurs actually behaving, how do we construct testable hypotheses about ancient behaviors? More specifically, how do we go about selecting from among various alternative explanations? At least three strategies are effective.

1) One strategy is to investigate the biomechanical properties of a given bony feature. For example, is the structure in question strong enough to withstand use as a weapon? In turns out that many bizarre structures—such as ceratopsid frills, hadrosaur crests, and stegosaur plates—are often thin and fragile, casting doubt on any presumed role as weapons or defense structures. In contrast, it has been argued[18] that the crests of at least some hadrosaurs had specialized acoustical properties, suggesting a potential role in vocalization.

2) Another approach is to reconstruct as accurately as possible the soft tissues, e.g., skin, nerves, blood vessels, muscles, attached to the bone.[19] For example, the presence of numerous bony grooves for blood vessels on the surface of *Stegosaurus* plates may suggest an abundant blood supply, which in turn might indicate a role in controlling body temperature.

3) The best source of information regarding dinosaur behavior typically comes from observations of behavior in living animals. These living animals include close relatives of dinosaurs (birds and crocodilians) that may share behaviors with dinosaurs by vir-

tue of shared ancestry, as well as distantly related animals that possess potentially analogous structures and behaviors.[15,16]

Various authors have argued that the predator defense and thermoregulation hypotheses are entirely inadequate to account for the sheer diversity in shapes and sizes of bizarre structures in dinosaurs, and that the mating signal hypothesis best accounts for this diversity. The most convincing evidence in support of this contention comes from modern analogies.[17] We see numerous examples of bizarre structures among living animals: peacock tails, lizard dewlaps, and deer antlers, among others. I regard it as compelling that among the many animals with horns or horn-like structures—from beetles and bison to chameleons and cassowaries—the primary function in virtually all cases involves the competition for mates. That is, bizarre structures are used first and foremost as mating signals to aid in reproductive success and the passing of one's genes to subsequent generations.

In support of the mate competition hypothesis, closely related species of horned and duckbill dinosaurs are distinguished overwhelmingly on the basis of skull roof features—horns and frills in ceratopsians (Fig. 2) and crests in hadrosaurs. While the rest of the skull and skeleton remained conservative in both groups, showing only minor evolutionary changes, each species is distinguished by its own decorative array of crests, hooks, horns, spikes, or other bony projections. Interestingly, pachycephalosaur domes, stegosaur plates and spikes, and ankylosaur tail clubs and armor are all primary characters used by paleontologists to identify species, suggesting a possible role in species recognition and mate competition.

Once again, this pattern holds for many kinds of modern animals. For example, closely related species of birds can often be distinguished solely on the color and shape of their plumage. This finding should not be surprising if we stop and think about it. The reason that paleontologists identify dinosaur species based on these weird features is that the dinosaurs themselves, like animals today, probably used the very same visual cues to recognize species members, as well as to compete for mates.

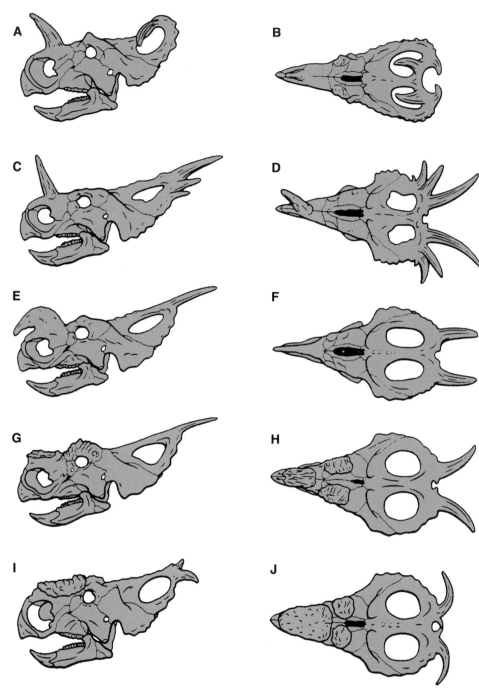

evolutionary role and predominant function. Based largely on evidence from living animals with bizarre structures, it seems most likely that at least the horns, frills, crests and other varied forms of headgear in dinosaurs functioned as mating signals. If so, they likely evolved to enhance opportunities for reproductive success through display and, in some cases, combat with members of the same species. With regard to hadrosaurs and ceratopsids, these features may indicate complex social organizations involving a hierarchy or "pecking order."

Dinosaurs as Failures

Dinosaurs are often regarded as the poster-children of failure, simply because they are both famous and dead. This portrayal is completely inappropriate for a number of reasons. First of all, dinosaurs persisted for 150 million years. They are one of the most successful animal groups ever to have existed on this planet. Although the comparison is inappropriate, since it deals with only a single species, the lifespan of *Homo sapiens* is a geological eyeblink, less than 300,000 years.

Second, extinction should not be accompanied by a black mark of failure. Throughout time, extinction has been the evolutionary rule, not the exception. It is estimated that over 99 percent of all species that have ever existed on this planet are now extinct.

Third, as discussed, dinosaurs did not actually go extinct 65 million years ago. They are

Fig. 2. Bizarre structures in the skulls of a group of horned dinosaurs, the centrosaurines, showing species-specific variations. Skulls are depicted in left lateral and dorsal views: **A, B.** *Centrosaurus apertus*; **C, D.** *Styracosaurus albertensis*; **E, F.** *Einiosaurus procurvicornis*; **G, H.** *Achelousaurus horneri*; **I, J.** *Pachyrhinosaurus canadensis*. From Sampson et al., 1997.

This is not to say that bizarre structures like horns and frills were never used for defense against predators or for aiding control of body temperature. These explanations are not mutually exclusive. The question is one of original

still with us today in the form of birds and, with over 9,000 living species, birds alone rank amongst the most successful of vertebrate groups.

Nonetheless, most of the dinosaurs are incontrovertibly gone and we want to know why. The dinosaur renaissance reinvigorated the study of dinosaurian lives and offered a subtly new perspective on their fate. Rather than being regarded as prehistoric monsters in evolution's backwater, lumbering through time awaiting the inevitable guillotine of extinction, paleontologists began re-envisioning dinosaurs as highly active, diverse and successful creatures, dominating terrestrial ecosystems worldwide for many millions of years. Emphasis has been placed more on how they lived than how they died, and this has led us to ask a very different question. Given their remarkable, widespread, and lengthy success, what combination of forces could possibly have brought an end to this dominating reign?

Dinosaurs lived during the Mesozoic Era, their fossilized remains occurring in rocks spanning 225 million years ago to about 65 million years ago. It has long been recognized that a major extinction punctuated the end of this lengthy tenure, extinguishing not only the dinosaurs but many other forms of life as well. The roster of fatalities includes flying reptiles (pterosaurs), sea-going reptiles (plesiosaurs and mosasaurs), the nautiloid-like ammonites, and various groups of plankton. Well over 100 ideas, ranging from reasonable to ridiculous, have been proposed to account for this most famous of dyings. Included in the long list of putative murderous causes are disease, slipped vertebral discs (because they were too big), loss of interest in sex, poison plants (leading either to diarrhea or constipation, depending on the author), climatic change in the form of global cooling or warming, cosmic radiation from a supernova, egg-eating mammals, sunspots, and (my personal favorite) trigger-happy aliens on interstellar hunting junkets.

By far the most commonly cited explanation today is the impact hypothesis.[20] This well-known idea postulated that an asteroid about 10 km in diameter collided with the Earth at a velocity of about 100,000 kph, re-

sulting in an almost inconceivable explosive force. Upon impact, the asteroid disintegrated, throwing smoke and dust high into the atmosphere, ultimately enveloping most of the world and transforming day to night. The end result was a prolonged period of cold and dark lasting several months. Photosynthesis would have halted due to lack of light, and acid rain would have destroyed much life. According to this view, dinosaurs and many other groups of animals perished as a direct result of the resulting environmental devastation.

This impact scenario is certainly spectacular, but is it correct? In the views of many, the actual chain of events may have been more complex. Abundant evidence now indicates that an extraterrestrial body did indeed slam into the planet 65 million years ago.[21] There is even a smoking gun, the Chicxulub Crater, near the Yucatan Peninsula in the Gulf of Mexico, which may preserve the site of impact. Nonetheless, some evidence suggests that the dinosaurs and various other groups may already have been on their way out. That is, the diversity of dinosaur species, in at least one area, appears to have decreased rather dramatically over the last few million years prior to 65 million years ago. The reasons for this drop in diversity, if it occurred at all, remain obscure, with potential contributory factors including environmental cooling and a major shift in vegetation regimes.

Rarely expressed, particularly in the popular media, is how little we actually know about this extinction event, particularly as it pertained to dinosaurs. Indeed, while it is widely assumed that all dinosaurs other than birds perished 65 million years ago, the evidence is relatively meager. That is, we still have very few examples of dinosaur remains right up to the extinction boundary. In fact, dinosaur-bearing boundary sediments have been examined in detail in only one place on Earth—eastern Montana. As debated previously,[22] it is possible that some dinosaurs persisted beyond the terminal Cretaceous boundary for a significant period.

It seems to me that the best places to look for post-Cretaceous dinosaurs will be isolated areas a great distance from the impact site,

which might have served as refuges for millions of years following the catastrophe of 65 million years ago. Personally, I hope that future fieldwork uncovers evidence of dinosaurs that lived long after the great extinction, revealing that this particular emperor of assumption has been naked all along.

In sum, several factors should be kept in mind when considering the dinosaurian demise. First, as with all true scientific hypotheses, any explanation must be testable. Second, any extinction hypothesis must account for as many lines of evidence as possible, e.g., the extinction of some plants and animals as well as the persistence of others. Third, in only a mere handful of areas on the globe (one in particular—eastern Montana) have dinosaur extinction ideas been tested, so there is plenty of room for new evidence and alternative hypotheses. Fourth, dinosaurs did not persist unchanged for 150 million years, only to be devastated in one tumultuous event. They underwent much evolution and turnover of species, and suffered other extinction events prior to 65 million years ago. The major point of departure between these earlier deaths and the mass extinction at the close of the Mesozoic is that no large-bodied dinosaur descendents replaced the extinct forms. And finally, recent work strongly supports the notion that dinosaurs did not truly go extinct. They are still with us today flying overhead. Rather than abject failures, dinosaurs are better regarded as an incredible success story worthy of celebration.

Dinosaurs, Education, and Integration

The history of science is not merely a saga of ever-increasing knowledge. It is better characterized as a lengthy series of passionate journeys. While the majority of these journeys lead ultimately to dead ends, some contribute kernels of insight, and an extremely small number fundamentally alter our understanding of the world. John Ostrom's discoveries and remarkable vision sparked a revolution in our perception of dinosaurs, impacting both scientists and the general public. This paradigm shift enabled us to see these long-dead beasts with new

eyes, and to explore previously unconceived questions. Happily, this time of discovery continues unabated. Dinosaur paleontologists have recovered and named more new species of dinosaurs during the past 25 years than in all of prior history, and there is no end in sight. New dinosaurs will be discovered generations from now and, more importantly, there will undoubtedly be periodic upheavals in our understanding of ancient life forms, as we ask new questions and apply new tools in the search for answers.

Traditionally, science was taught as a body of facts documenting the way the world works. Consequently, many children came to regard biology, physics, math, and other areas of science as tedious amalgams of theorems, laws, and arcane methodologies. Not surprisingly, this cold, even cynical perspective often persisted through adulthood, accompanied by an overriding belief that little remains to be discovered. In contrast to this view, we are always in a time of discovery, not just in paleontology but in all areas of science. Whether one's interests lie in paleontology, chemistry, biology, geology or some other field, there are still many, many lifetimes worth of new ideas out there. Put another way, the communication of science should not revolve solely around what we think we know, but should also emphasize all that we don't know. In this way, we can engender and nurture the spirit of discovery that drives most scientists, and carry this spirit like a torch to illuminate the pathway for others.

Given the inherent interdisciplinary nature of paleontology, dinosaurs offer an ideal venue to teach an integrative, systems-based perspective. Indeed dinosaurs provide an exceptional access point for communicating virtually any area of science—from physics, chemistry, and mathematics to geology, ecology, and climatology. For example, a look at the cold-blooded/warm-blooded debate enables one to address a broad range of topics including physiology, behavior, ecology, even climate and plate tectonics (with regard to dinosaurs living near the poles). Moreover, as much as any other topic in science, paleontology has a broad allure and accessibility that transcends generations. Indeed, for many children, dinosaurs represent the first

foray into science. In short, dinosaurs are not only fascinating in and of themselves. With a little creativity and background, they also offer numerous exciting opportunities to illustrate the intricate networks that interconnect all physical and living systems.

R e f e r e n c e s

1. Kuhn, T. 1970. *The structure of scientific revolutions*. Chicago: University of Chicago.

2. Ostrom, J.H. 1969. Osteology of *Deinonychus antirrhopus*, an unusual theropod from the Lower Cretaceous of Montana. *Bulletin of the Peabody Museum of Natural History* 30:1–165.

3. Ostrom, J.H. 1973. The ancestry of birds. *Nature* 242(5393):136.

4. Ostrom, J.H. 1976. *Archaeopteryx* and the origin of birds. *Biological Journal of the Linnean Society* 8(2):91–182.

5. Padian, K., and L.M. Chiappe. 1998. The origin and early evolution of birds. *Biological Reviews* 73:1–42.

6. Xu, X., Z. Tang, and X. Wang. 1999. A therizinosauroid dinosaur with integumentary structures from China. *Nature* 399:350–354.

7. Bakker, R.T. 1975. Dinosaur renaissance. *Scientific American* 232(4):55–78.

8. Carrano, M.T. 1999. What if anything is a cursor? Categories versus continua for determining locomotor habit in mammals and dinosaurs. *Journal of Zoology, London* 247:29–42.

9. Thomas, R.D.K. and E.C. Olson (eds.). 1980. *A cold look at the warm-blood dinosaurs*. American Association for the Advancement of Science Selected Symposium no. 28.

10. Horner, J.R., A. de Ricqles, and K. Padian. 1999. Variation in dinosaur skeletochronology indicators: implications for age assessment and physiology. *Paleobiology* 25(3):295–304.

11. Reid, R.E.H. 1997. The case for "intermediate" dinosaurs. In *The complete dinosaur*, eds. J.O. Farlow and M.K. Brett-Surman, 449–473. Bloomington: Indiana University Press.

12. Farlow, J.O., C.V. Thompson, and D.E. Rosner. 1976. Plates of *Stegosaurus*: forced convection or heat loss fins? *Science* 192:1123–1125.

13. Wheeler, P.E. 1978. Elaborate CNS cooling structures in large dinosaurs. *Nature* 275:441–443.

14. Davitashvilli, L.S. 1961. *Teoriya polovogo otbora (The theory of sexual selection)*. Moscow: Izdatel'stvo Akademii Nauk (Acad. of Sciences Press).

15. Farlow, J.O., and P. Dodson. 1975. The behavioral significance of frill and horn morphology in ceratopsian dinosaurs. *Evolution* 29:353–361.

16. Hopson, J.A. 1975. The evolution of cranial display structures in hadrosaurian dinosaurs. *Paleobiology* 1:21–43.

17. Sampson, S.D., M.J. Ryan, and D.H. Tanke. 1997. Craniofacial ontogeny in centrosaurine dinosaurs (Ornithischia: Ceratopsidae): taxonomic and behavioral implications. *Zoological Journal of the Linnean Society* 221(2):293–337.

18. Weishampel, D.B. 1981. Acoustic analyses of potential vocalizations in lambeosaurine dinosaurs (Reptilia: Ornithischia): comparative anatomy and homologies. *Journal of Paleontology* 55:1046–1057.

19. Witmer, L.M. 1995. The extant phylogenetic bracket and the importance of reconstructing soft tissues in fossils. In *Functional morphology in vertebrate paleontology*, ed. J. Thomason, 19–33. New York: Cambridge University Press.

20. Alvarez, W., and F. Azaro. 1990. An extraterrestrial impact. *Scientific American* 263(4):44–52.

21. Archibald, J.D. 1996. *Dinosaur extinction and the end of an era: what the fossils say*. New York: Columbia University Press.

22. Rigby, J.K., Jr. 1987. The last of the North American dinosaurs. In *Dinosaurs past and present*, vol. II, eds. S.J. Czerkas and E.C. Olson, 119–135. Seattle: University of Washington Press.

23. Barnett, L. 1955. *The world we live in*. New York: Time Inc.

24. Paul, G.S. 1988. *Predatory dinosaurs of the world*. New York: Simon and Schuster.

25. Czerkas, S.A. 1986. A reevaluation of the plate arrangement on *Stegosaurus stenops*. In *Dinosaurs past and present*, vol. II, eds. S.J. Czerkas and E.C. Olson, 83–99. Seattle: University of Washington Press.

Acknowledgments

I cordially thank Mark Loewen, Judy Scotchmoor, and two anonymous reviewers for providing valuable comments on an earlier draft of this paper. Figure 1 was executed by Jude Higgins; figure 2 by Ed Heck.

Dinosaurs: A Time and Place

The earliest history of the dinosaurs, their evolution and diversification, is the focus of the article by **Martin**. He explains how paleontologists use numerous lines of evidence to hypothesize about the physical features that were present in "the first dinosaur," and why our hypotheses about dinosaurs can change as new fossil material is uncovered. He provides a physical and temporal framework within which to understand possible causes for adaptation and diversification of dinosaurs through the Mesozoic.

Understanding how scientists classify organisms can be confusing to those unfamiliar with the methods used. **Holtz**'s article helps demystify the process. He takes readers through a brief history of classification schemes, and explains how the modern method called "cladistics" uses the presence or absence of particular inherited features (such as bones, scales, or fur) to place organisms into categories according to their evolutionary histories. In the process, he illustrates how paleontologists test the results of cladistic analysis to arrive at the most probable family tree of dinosaurs.

Explaining how geologists unravel the concept of geologic time, **Lucas** puts dinosaurs into a geologic timeframe. He describes the process of defining relative and numerical time units—for example, how we decide when the Triassic Period began, or when the Mesozoic Era ended—and how we can say that certain rocks in widely separated parts of the world belong to the "Age of Dinosaurs."

The contribution by **Forster** describes how paleontologists know where dinosaurs lived and why not every Mesozoic rock unit yields dinosaur fossils. She explains how scientists use knowledge of dinosaur distributions to learn more about dinosaur evolution and lifestyles, as well as how putting dinosaurs in the context of their paleoenvironments and paleogeography enables scientists to ask and answer even more questions about the history of life on Earth.

All living organisms must eat, excrete, respire, protect themselves, and reproduce if the species is to continue. It is scientifically a pretty safe assumption that dinosaurs were no different. But, how can we know what they ate, how they moved, or how they reproduced? **Fastovsky** gives readers a view into the lost lives of dinosaurs. By examining the types of evidence paleontologists use to construct hypotheses about the paleoecology of dinosaurs, he describes the keys to understanding their world.

Dinosaurs were not the only animals roaming the world during the Mesozoic. **Fraser**'s article introduces readers to some of the rest of the cast of characters—large and small, vertebrate and invertebrate—that shared the Earth with the "terrible lizards." In the process, he helps clear up some misconceptions about what is and is not a "dinosaur:" *Triceratops*, yes…*Pteranodon*, *Dimetrodon*, and icthyosaurs, not!

Dinosaur Evolution: From Where Did They Come and Where Did They Go?

3

Anthony J. Martin

Department of Environmental Studies
Emory University, Atlanta, Georgia

The rhetorical and two-part question "From where did dinosaurs come and where did they go?" encourages the asking of many more questions, but such questions will inevitably relate to two intertwined phenomena experienced by all life: evolution and extinction. Dinosaurs, like all living things, originated through evolutionary processes, diversified as a result of those very same processes, and most of their lineages went extinct. In this chapter, I will focus on the processes and products of evolution in dinosaurs, but with emphasis on the beginning of that history. Hypotheses for dinosaur extinctions 65 million years ago and the evolution of some dinosaur lineages into birds are discussed in detail by other authors in this volume (see Currie, page 89; and Archibald, page 99).

Scientific studies of dinosaur evolution exemplify that science is a work in progress. Its practitioners often acknowledge that once-stated certainties are liable to be proven wrong, once-outlandish ideas may be closer to reality than originally thought, or, alternatively, only minor tweaking is needed to clarify some previously accepted hypotheses. For example, the statement "Dinosaurs evolved from reptiles in the latter part of the Late Triassic Period and went extinct at the end of the Cretaceous Period" has been amended in the past 20 years to "Dinosaurs evolved from archosaur lineages in the earliest part of the Late Triassic and some of them are still with us today as birds." Such shifts illustrate how scientists revise or disprove hypotheses more than affirm them (and often are delighted when they do so!), and dinosaurs in particular constitute excellent subjects for revisionist thinking. Each new discovery of dinosaur remains or dinosaur trace fossils provides yet more evidence for testing whether previously accepted concepts about dinosaurs and their evolution (especially their origins) are still justified.

> *Each new discovery . . . provides yet more evidence for testing whether previously accepted concepts about dinosaurs and their evolution (especially their origins) are still justified.*
>
> Tony Martin

Anthony "Tony" J. Martin is a senior lecturer in the Department of Environmental Studies at Emory University in Atlanta, Georgia. He received his Ph.D. from the University of Georgia. Martin teaches undergraduate classes in historical geology, environmental geology, and environmental science, as well as field courses on dinosaurs, desert geology, and tropical environments. He has presented more than 40 papers at regional, national, and international meetings on his main research interest, ichnology, the study of modern and fossil traces. He recently wrote a textbook on dinosaurs intended for undergraduate non-science majors *Introduction to the Study of Dinosaurs* (2001, Blackwell Science).

Dinosaur Ancestors and the Origins of Dinosaurs

What were the most likely ancestors of dinosaurs? Intrinsic to this question is the dilemma facing a paleontologist, which is how to define the "first dinosaur" from the fossil record. For example, *Stegosaurus stenops* of the Late Jurassic Period and *Triceratops horridus* of the Late Cretaceous were certainly dinosaurs (just ask any

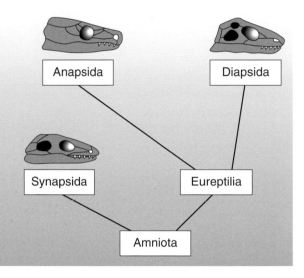

Fig. 1.

Synapsid, anapsid, and diapsid skull types in a cladogram showing their hypothesized evolutionary relationships. Temporal fenestra indicated by blackened areas; orbits (eye sockets) are shaded. Adapted from Martin (2001: Fig. 10.5).

five-year-old), but what about *Eoraptor lunensis* and *Herrerasaurus ischigualasto* of the Late Triassic? Some paleontologists think that *Eoraptor* was a dinosaur,[1, 2] whereas others do not,[3] and similar dissent is expressed about *Herrerasaurus*.[4] Additionally, given the transitional nature of evolution, any declaration made by a paleontologist that he or she found evidence for the "first dinosaur" might be analogous to stating that a shade of gray is exactly between white and black. The fossil evidence for the evolutionary lineage for dinosaurs, particularly in their early history, is not well known. Does this situation mean that no hypotheses about the origins of the first dinosaurs can be made? Of course not—paleontologists can still hypothesize, because after all, a little bit of evidence is better than none. Moreover, paleontologists acknowledge that although they can only hypothesize on the basis of what evidence they have now, they also optimistically remind themselves that the fossil record gets better every day as more discoveries are made

and studied. As a result, a definition of the "first dinosaur" can be given here, but with the understanding that undoubtedly it will be refined and clarified later.

Origin of Dinosaur Ancestors

The ancestors and origin of dinosaurs could be traced back to the origin of life itself. However, for the sake of brevity, let us begin with amniotes—four-legged vertebrates (tetrapods) that evolved enclosed (cleidoic) eggs for reproduction. Amniotes appeared by the beginning of the Pennsylvanian Period, about 310 million years ago.[5, 6] The evolution of a cleidoic egg was a huge step for tetrapods because it allowed them to inhabit areas that were dry or distant from bodies of water. No longer were tetrapods bound to certain terrestrial ecosystems. One of the predictions of evolutionary theory is that they should have undergone relatively rapid diversification as a reflection of adapting to these different ecosystems. Indeed, this was the case, and this diversification can best be seen by looking at changes in their skulls and limbs through the latter part of the Paleozoic Era.

Amniotes are broadly divided into three evolutionarily related groups—Anapsida, Synapsida, and Diapsida, based on the number of temporal fenestrae they possess (Fig. 1). Place a couple of fingers behind your eye socket on the side of your skull. You will feel a soft area there. Your hand is feeling a temporal fenestra, which is a hole in the structure of your skull caused by your cheekbone: the cheekbone flares out around the fenestra. Anapsids lack temporal fenestrae; synapsids have one on each side of the skull; and diapsids have two on each side of the skull. (Using this definition, you are a synapsid.) Based on current data, diapsids and anapsids probably evolved from a common ancestor, hence they are put in the same group (Eureptilia) and separated from synapsids.

The diapsid group, or clade, is where we will find dinosaurs. During the later part of the Permian Period (about 260 million years ago), diapsids diverged into two clades: Lepidosauria (lizards and snakes) and Archosauria. Archo-

saurs are distinguished by the following traits: openings in the skull in front of the eye sockets, laterally compressed and serrated teeth, holes in the front of the lower jaw (dentary), modified ankle bones, and longer bones in the pelvis.[7] Archosaurs show up in the fossil record in the Early Triassic and evolved into crocodiles, pterosaurs, dinosaurs, and birds. A fossil that is a likely candidate for a common ancestor to these archosaurs is a one-meter long animal—*Euparkeria*, of the Early Triassic of South Africa.[8]

Archosaurs then diverged into two clades during the Triassic Period: Pseudosuchia, which included many fossil forms, as well as the living crocodilians,[9] and Ornithodira. Dinosaurs are ornithodirans, originating from that clade either during the latest part of the Middle Triassic or the earliest part of the Late Triassic, about 230 million years ago. Although pterosaurs are not "flying dinosaurs," they are also ornithodirans, and thus are closely related to dinosaurs; that is, they share a common ancestor.

First Dinosaurs:
What, When, and Where?

So what did the "mother of all dinosaurs" look like? Based upon features common to the earliest dinosaurs and pterosaurs, this ancestor would have been bipedal, had hind limbs longer that its forelimbs, had four (or five) digits on its hand, long metatarsals and digits on its foot, a uniquely hinged ankle, and a tibia

and fibula longer than its femur.[10, 11] Thus far, the best fits to this description are represented by *Marasuchus* (synonymized with *Lagosuchus* by some paleontologists) and *Lagerpeton*, which both occur in Middle Triassic strata of Argentina.[12, 13] *Marasuchus* and *Lagerpeton* apparently originated very close to the divergence time of ornithodirans, meaning that their descendants could have become either pterosaurs or dinosaurs.

Three fossils from the earliest part of the Late Triassic (~227.8 mya) can be considered candidates for the oldest known dinosaur (Figs. 2, 3): *Eoraptor lunensis* and *Herrerasaurus ischigualasto* from the Ischigualasto Formation of Argentina, and *Staurikosaurus pricei* from the Santa Maria Formation of Brazil.[14, 6, 17, 1, 18, 19, 2] Of these species, the one that causes the most arguments among paleontologists is *Eoraptor*. Its primitive traits, such as the lack of a joint within its mandible, as well as hindlimb and pelvic bones that differ slightly from those of all other dinosaurs,[2] prompted skepticism as to whether it is indeed a dinosaur.[3] Whichever

may be the best candidate, most paleontologists agree that these "first" dinosaurs were saurischians ("lizard-hipped" dinosaurs) and that they were evolutionarily connected with theropods.[20, 21, 22] However, the exact evolutionary relationships with ancestor and descendant species are still the subject of debate.[23]

Fig. 2. Profiles of three Late Triassic ornithodirans defined as primitive dinosaurs. **A.** *Eoraptor lunensis*. **B.** *Staurikosaurus pricei*. **C.** *Herrerasaurus ischigualasto*. Adapted from Martin (2001: Fig. 10.10). Bar scale = 1 m.

Fig. 3. *Herrerasaurus ischigualasto* of the Late Triassic of Argentina, interpreted as one of the first dinosaurs. Specimen in The Field Museum, Chicago, IL.

Competing with these fossils for the title of "oldest known dinosaur" are a few recent finds of prosauropod fragments in Madagascar, which some paleontologists assert were in strata older (230–235 mya) than those of *Eoraptor* and *Herrerasaurus*.[24] Unfortunately, no radiometric ages are yet calculated for the Madagascar rocks and the age estimate, which dips into the Middle Triassic, is currently based only on associated fossils. Nevertheless, this exciting find has helped to show how the beginning of dinosaur evolution is likely to be pushed back slightly in geologic time.

Supplementary evidence for the advent of dinosaurs comes not from skeletal evidence, but from tracks. Some tracks in the latest part of the Middle Triassic and earliest part of the Late Triassic closely resemble what most paleontologists agree are dinosaur tracks.[25, 26] Because the anatomy of probable dinosaur ancestors points toward bipedal walking and a foot with three prominent toes, their tracks should reflect this two-legged walking and a three-toed compression shape. Considering the large number of tracks left by a living and moving animal in comparison to its single body, early dinosaur tracks may be more common than their fossilized bodies. However, skepticism about what animals made what tracks is necessary and good skeletal data correlations are needed before making any conclusions about dinosaurs walking in the Middle Triassic.[26]

Early Dinosaur Diversification

Finds of primitive dinosaurs and their probable ancestors in South America, in addition to those recently found in Madagascar, currently point toward an origin of dinosaurs from the southern continent of Gondwana during the Middle to Late Triassic. How they diversified immediately after their origin is a good question. One hypothesis is that early dinosaurs successfully competed with other archosaurs for habitats and resources throughout the latter part of the Triassic, which caused some pseudosuchians to go extinct and ornithodirans to endure.[27] However, this hypothesis is not very well supported by the fossil evidence, in that inverse trends between dinosaur abundance

and extinctions of other archosaurs are not clear. Indeed, some archosaur lineages show signs of going extinct well before dinosaurs were established, suggesting that environmental factors were primarily affecting their natural selection, rather than interspecific competition with dinosaur ancestors.[28, 29] The latest addition to this hypothesis invokes a meteorite-linked extinction of pseudosuchians that opened the way for the 140-million-year dinosaurian hegemony that began in the last part of the Triassic.[30]

For a broad overview of how a changing world may have contributed to the development of dinosaurs, we must look at what was happening during the Late Triassic. About 220 million years ago, the supercontinent Pangea began to break up, resulting in major sea level fluctuations.[31] Evidence of environmental change and its effects on flora and fauna during the Late Triassic includes major extinctions of marine invertebrates. The splitting of Pangea no doubt triggered changes in the ocean and atmosphere, which in turn would have altered global climate. Geologic evidence of such climate change includes thick and widespread Triassic evaporite deposits, formed by the evaporation of seawater in restricted basins during extended times of aridity.[32, 33, 34, 35]

The arid Triassic climates impacted terrestrial communities. Selection favored drought-resistant plants, which became dominant during this time. With a change in the availability of food sources to herbivorous tetrapods, picky eaters would have gone extinct and generalists would have survived to spread their genes to a new generation. Is this mere speculation? No, the fossil record of plants from the Late Triassic shows such a shift in correspondence with climate change,[36, 37] and this change also roughly correlates with changes in terrestrial archosaur assemblages. Of course, correlation is not necessarily causation, thus much more data are needed before making a more definitive linkage of climatic and ecologic factors with pseudosuchian extinctions and dinosaur evolution. Nevertheless, this evidence provides a starting point for review or revision of such hypotheses.

The breakup of Pangea, in combination with sea-level fluctuations, also caused habitat

fragmentation on a massive scale, meaning that dinosaur and other archosaur faunas became geographically isolated. Such a situation would have been conducive for speciation as isolated populations adapted to their new ecosystems. This scenario is also backed by geologic and paleontologic evidence. Diversification in the fossil record shows an approximate correlation with continental breakups during the Phanerozoic Eon, long before the Mesozoic.[38] One of the hypotheses to explain this diversification is that new ecosystems provided new niches for species with the genetic ability to adapt. Dinosaurs thus would have had new adaptations in the Late Triassic that likely interrelated with newly available niches caused by both continental rifting and the abandonment of niches as other archosaurs went extinct.[39]

As mentioned earlier, a new revelation contrasting with (or augmenting) the preceding gradual scenario concerns a possible meteorite impact near the end of the Triassic Period. The evidence cited for this hypothesis is multifaceted, drawn from archosaur body fossil assemblages, dinosaur tracks, iridium concentrations, and fern spores in sediments spanning the Triassic-Jurassic boundary.[30] The gist of this hypothesis is that pseudosuchians already had depleted numbers when a meteorite (sufficiently large enough to cause ecological damage) struck the Earth. Then they quickly became extinct. Meanwhile, ferns, as typical "first colonizers" after an ecological catastrophe, became more abundant after the impact, and the surviving ornithodirans (including dinosaurs) were left with ecological niches newly unoccupied by other archosaurs. The idea of a meteorite impact near or at the end of the Triassic that had an adverse effect on global ecosystems is not new,[40, 41] but the subsequent refining of this hypothesis illustrates a progress of the science behind it.

Early Evolutionary Trends in Dinosaurs

Dominance in Size and Number

Regardless of the large- and small-scale evolutionary factors that shaped dinosaur evolution, their ascent to being the dominant land animals of terrestrial ecosystems within only about 25 million years is certainly noteworthy. The overall evolutionary trends observed in dinosaurs after their early evolutionary history include an increase in larger body sizes (especially in theropods and sauropodomorphs), increased diversity (i.e., more species), and a higher representation within terrestrial faunas. As represented by the skeletal record, dinosaurs went from about 6 percent of terrestrial amniote species to as much as 60 percent by the end of the Triassic Period.[42] The switch from relatively uncommon and small theropods to abundant and much larger theropods toward the end of the Triassic is also well reflected by dinosaur tracks.[43, 30]

Changes in Body Plans

More specific evolutionary trends include changes in other aspects of body plans, such as how some prosauropods (e.g., *Plateosaurus*), became adapted to a quadrupedal lifestyle, despite all indications that the most immediate ancestors of dinosaurs were bipedal.[44] This change in locomotion was likely a consequence of increased body size in prosauropods toward the end of the Triassic, in which they became the largest land herbivores of that time.

A Shift to Herbivory

Dinosaurs showed rapid diversification with respect to feeding, considering that carnivorous and herbivorous dinosaurs show up at nearly the same time in the geologic record. This early shift to herbivory for dinosaurs can be illustrated by prosauropod teeth from the Late Triassic, teeth that must have evolved in accordance with the availability of certain plants as food.[17] Prosauropod teeth were certainly a departure from the pointed and blade-like teeth of early theropods that were used for meat-eating. Prosauropod herbivory was apparently augmented by the use of gastroliths ("stomach stones") that were probably used to grind tough-to-digest food.[45, 46, 47] The use of gastroliths by terrestrial vertebrates for processing plant material was rare until the evolution of herbivorous dinosaurs. Throughout the remainder of the Mesozoic, sauropodomorphs, in particular, used them as aids to digestion.[48, 49]

Parenting

Reproductive behaviors may have been similar to those of later generations of dinosaurs, but only two Late Triassic dinosaur nests (containing dinosaur egg clutches) have been found so far.[50, 51] Thus, much is still unknown about whether dinosaur parents brooded or cared for their young, behaviors proposed for some Late Cretaceous dinosaurs (see Horner, page 71).[52]

Thermoregulation

Changes in thermoregulation ("cold-blooded" versus "warm-blooded") in archosaur lineages leading to dinosaurs are likewise poorly understood, but remain as intriguing areas for future research, considering current controversies about dinosaur thermoregulation (see de Ricqlès, page 79).[53, 54]

The fact that dinosaurs survived environmental factors that caused the extinctions of other archosaurs means that dinosaur ancestors had genes that favored their selection. These genes were later modified considerably during the successive 140-million years of the Jurassic and Cretaceous periods, but the persistence and diversification of dinosaurs during such a long span of time bespeaks of the successes of their initial adaptations. Their long story, still incomplete, began with their seemingly humble and obscure origins in the Triassic.

Where Did Dinosaurs Go?

Any cladistically inclined paleontologist will gleefully inform you that dinosaurs never went extinct, they are still here as birds. For those who do not know the "secret handshake" shared by cladists, one might think that such pronouncements are attributable to an unquestioning acceptance of what is printed in tabloid newspapers. As odd as the statement "birds are dinosaurs" may sound, it is just as apt as saying "humans are mammals." The key to understanding why birds are dinosaurs is related to hypotheses concerning ancestor-descendant relationships, just as was demonstrated for archosaurian ancestors of dinosaurs. Thus far, the hypothesis that modern birds descended from theropod ancestors in the Mesozoic Era (likely in the Middle or Late Jurassic) has not been falsified, despite its thorough testing for nearly 140 years. This line of inquiry into bird origins with relation to dinosaurs began when a skeleton of *Archaeopteryx* was discovered in 1861 and was immediately noted as an intermediate form between reptiles and birds by naturalist T.H. Huxley (see Currie, page 89).[55]

But to tell anything more about this story would be preemptory. Rest assured that a combination of genetic and environmental factors were involved, and we can observe those factors today. Birds represent some of the best examples of animals we have actually watched evolve (such as the famous "Darwin's finches" of the Galapagos [56, 57]). As a result, data pertinent to the evolution of dinosaurs will continue to be collected as birds and other archosaurs (such as crocodiles) are studied today, and given enough time, major parts of this chapter will have to be amended. The foundation of evolution is change, and the same goes for our knowledge of evolution. In this sense, the evolution of dinosaurs is no exception.

References

1. Sereno, P.C., and F.E. Novas. 1992. The complete skull and skeleton of an early dinosaur. *Science* 258:1137–1140.

2. Sereno, P.C., C.A. Forster, R.R. Rogers, and A.M. Monetta. 1993. Primitive dinosaur skeleton from Argentina and the early evolution of Dinosauria. *Nature* 361:64–66.

3. Padian, K., and C.L. May. 1993. The earliest dinosaurs. *New Mexico Museum Natural History Science Bulletin* 3:379–381.

4. Novas, F.E. 1997. Herrerasauridae. In *Encyclopedia of dinosaurs*, eds. P.J. Currie and K. Padian. San Diego: Academic Press.

5. Carroll, R.L. 1988. *Vertebrate Paleontology and Evolution*. New York: W.H. Freeman.

6. Colbert, E.H., and M. Morales. 1991. *Evolution of the vertebrates: a history of the backboned animals through time*. New York: John Wiley and Sons.

7. Parrish, J.M. 1997. Evolution of the archosauria. In *The complete dinosaur*, eds. J.O. Farlow and M.K. Brett-Surman, 191–203. Bloomington: Indiana University Press.

8. Ewer, R.F. 1965. The anatomy of the thecodont reptile *Euparkeria capensis* Broom. *Philosophical Transactions of Royal Society London B* 248:379–435.

9. Holtz, T.R., Jr. 2000. Classification and evolution of dinosaur groups. In *The Scientific American book of dinosaurs*, ed. G.S. Paul, 140–168. St. Martin's Press.

10. Cruickshank, A.R.I., and M.J. Benton. 1985. Archosaur ankles and the relationships of the thecodontian and dinosaurian reptiles. *Nature* 317:715–717.

11. Padian, K., and K.D. Angielczyk. 1999. Are there transitional forms in the fossil record? In *The evolution-creation controversy II: perspectives on science, religion, and geological education*, eds. P.H. Kelley, J.R. Bryan, and T.A. Hansen, T.A. The Paleontological Society Papers 5:47–82.

12. Bonaparte, J.F. 1975. Nuevos materiales de *Lagosuchus talampayensis* Romer (Thecodontia-Psuedosuchia) y su significado en el origen de los Saurischia. Chañarense inferior, Triásico medio de Argentina. *Acta Geologica Lilloana* 13:5–90.

13. Sereno, P.C., and A.B. Arucci. 1993. Dinosaur precursors from the Middle Triassic of Argentina: *Lagerpeton chanarensis*. *Journal of Vertebrate Paleontology* 13:385–399.

14. Reig, O.A. 1963. La presencia de dinosaurios saurisquios en los "Estratos de Ischigualasto" (Mesotriásico superior) de las provincias de San Juan y La Rioja (República Argentina). *Ameghiniana* 3:3–20.

15. Colbert, E.H. 1970. A saurischian dinosaur from the Triassic of Brazil. *American Museum Novitates* 2405:1–9.

16. Galton, P.M. 1977. On *Staurikosaurus pricei*, an early saurischian dinosaur from Brazil, with notes on the Herrerasauridae and Poposauridae. *Paläontologisch Zeitschrift* 51:234–245.

17. Galton, P.M. 1986. Herbivorous adaptations of Late Triassic and Early Jurassic dinosaurs. In *The beginning of the age of dinosaurs*, ed. K. Padian, 203–221. Cambridge: Cambridge University Press.

18. Novas, F.E. 1992. Phylogenetic relationships of the basal dinosaurs, the Herrerasauridae. *Palaeontology* 35:51–62.

19. Novas, F.E. 1993. New information on the systematics and postcranial skeleton of *Herrerasaurus ischigualastensis* (Theropods: Herrerasauridae) from the Ischigualasto Formation (Upper Triassic) of Argentina. *Journal of Vertebrate Paleontology* 13:400–423.

20. Sereno, P.C. 1993. The pectoral girdle and forelimb of the basal theropod *Herrerasaurus ischigualastensis*. *Journal of Vertebrate Paleontology* 13(4):425–450.

21. Sereno, P.C., and F.E. Novas. 1993. The skull and neck of the basal theropod *Herrerasaurus ischigualastensis*. *Journal of Vertebrate Paleontology* 13(4):451–476.

22. Hunt, A.P., S.G. Lucas, A.B. Heckert, R.M. Sullivan, and M.G. Lockley. 1998. Late Triassic dinosaurs from the western United States. *Geobios* 31:511–531.

23. Lucas, S.G., A.P. Hunt, and R.A. Long. 1992. The oldest dinosaurs. *Naturwissenschaften* 79:171–172.

24. Flynn, J.J., R.L. Whatley, A.R. Wyss, J.M. Parrish, B. Rakotosamimanana, and W.F. Simpson. 1999. A Triassic fauna from Madagascar, including early dinosaurs. *Science* 286:763–765.

25. Demathieu, G.R. 1989. Appearance of the first dinosaur tracks in the French Middle Triassic and their probable significance. In *Dinosaur tracks and traces*, eds. D.D. Gillette and M. Lockley, 201–207. Cambridge: Cambridge University Press.

26. King, M.J., and M.J. Benton. 1996. Dinosaurs in the Early and Middle Triassic?—the footprint evidence from Britain. *Palaeogeography, Palaeoclimatology, Palaeoecology* 122:213–225.

27. Charig, A. 1984. Competition between therapsids and archosaurs during the Triassic Period: a review and synthesis of current series. *Symposia of the Zoological Society of London* 52:597–628.

28. Benton, M.J. 1983. Dinosaur success in the Triassic: a noncompetitive ecological model. *Quarterly Review of Biology* 58:29–55.

29. Benton, M.J. 1990. Origin and interrelationships of dinosaurs. In *The dinosauria*, eds. D.B. Weishampel, P. Dodson, and H. Osmólska, 11–30. Berkeley: University of California Press.

30. Olsen, P.E., D.V. Kent, H.-D. Sues, C. Koeberl, H. Huber, A. Montanari, E.C.A. Rainforth, S.J. Fowell, M.J. Szajna, and B.W. Hartline. 2002. Ascent of dinosaurs linked to an iridium anomaly at the Triassic-Jurassic boundary. *Science* 296:1305–1307.

31. McRoberts, C.A., C.R. Newton, and A. Allasinaz. 1995. End-Triassic bivalve extinction: Lombardian Alps, Italy. *Historical Biology* 9:297–317.

32. Hallam, A. 1985. A review of Mesozoic climates. *Journal of the Geological Society* (London) 142:433–445.

33. Parrish, J.T. 1993. Climate of the supercontinent Pangea. *Journal of Geology* 101:215–233

34. Pollard, D., and M. Schulz. 1994. A model for the potential locations of Triassic evaporite basins driven by paleoclimatic GCM simulations. *Global and Planetary Change* 9:233–249.

35. El-Tabakh, M., R. Riccioni, and B.C. Schreiber. 1997. Evolution of late Triassic rift basin evaporites (Passaic Formation): Newark Basin, eastern North America. *Sedimentology* 44:767–790.

36. Simms, M.J., and A.H. Ruffell. 1990. Climatic and biotic change in the Late Triassic. *Journal of Geological Society of London* 147:321–327.

37. Ziegler, A.M., J.M. Parrish, J. Yao, E.D. Gyllenhaal, D.B. Rowley, J.T. Parrish, S. Nie, A. Bekker, and M.L. Hulver. 1993 . Early Mesozoic phytogeography and climate. *Philosophical Transactions of the Royal Society of London B* 341:297–305.

38. Valentine, J.W., and E.M. Moores. 1972. Global tectonics and the fossil record. *Journal of Geology* 80:167–184.

39. Benton, M.J. 1986. The Late Triassic tetrapod extinction events. In *The beginning of the age of dinosaurs*, ed. K. Padian, 303–320. Cambridge: Cambridge University Press.

40. Olsen, P.E., N.H. Shubin, and M.H. Anders. 1987. New Early Jurassic tetrapod assemblages constrain Triassic-Jurassic tetrapod extinction event. *Science* 237:1025–1029.

41. Bice, D.M., C.R. Newton, S. McCauley, P.W. Reiners, and C.A. McRoberts. 1992. Shocked quartz at the Triassic-Jurassic boundary in Italy. *Science* 255:443–446.

42. Benton, M.J. 1993. Late Triassic extinctions and the origin of the dinosaurs. *Science* 260:769–770.

43. Lockley, M.G., and A.P. Hunt. 1995. *Dinosaur tracks and other fossil footprints of the western United States*. New York: Columbia University Press.

44. Christian, A., and H. Preuschott. 1996. Deducing the body posture of extinct large vertebrates from the shape of the vertebral column. *Palaeontology* 39:801–812.

45. von Huene, F. 1932. Die fossile Reptile-Ordnung Saurischia, ihre Entwicklung und Geschichte. Monogr. Geol. *Palaeontol.* (Parts I and II) 4:1–361.

46. Galton, P.M. 1973. On the anatomy and relationships of *Efraasia diagnostica* (v. Huene) n. gen., a new prosauropod dinosaur (Reptilia: Saurischia) from the Upper Triassic of Germany. *Paläontologische Zeitschrift* 47:229–255.

47. Raath, M. 1974. Further evidence of gastroliths in prosauropod dinosaurs. *Arnoldia* 7:1–5.

48. Calvo, J.O. 1994. Gastroliths in sauropod dinosaurs. *Gaia* 10:205–208.

49. Gillette, D.D. 1994. *Seismosaurus, the Earth shaker*. New York: Columbia University Press.

50. Bonaparte, J.F., and M. Vincent. 1979. El hallazgo del primer nido de dinosaurios triásicos (Saurischia Prosauropoda), Triásico superior de Patagonia, Argentina. *Ameghiana* 16:173–182.

51. Kitching, J.W. 1979. Preliminary report on a clutch of six dinosaurian eggs from the Upper Triassic Elliot Formation, Northern Orange Free State. *Paleontographica Africana* 22:41–45.

52. Horner, J.R. 2000. Dinosaur reproduction and parenting. *Annual Review of Earth and Planetary Sciences* 28:19–45.

53. Ruben, J.A., T.D. Jones, P.J. Currie, J.R. Horner, G. Espe, III, W.J. Hillenius, N.R. Geist, and A. Leitch. 1996. The metabolic status of some Late Cretaceous dinosaurs. *Science* 273:1204–1207.

54. O'Connor, M.P., and P. Dodson. 1999. Biophysical constraints on the thermal ecology of dinosaurs. *Paleobiology* 25:341–368.

55. Huxley, T.H. 1868. On the animals which are most nearly intermediate between birds and reptiles. *American Magazine of Natural History* 4:66–75.

56. Darwin, C.R. 1839. *Journal of researches into the geology and natural history of the various countries visited by the H.M.S. Beagle*. London: Henry Colburn.

57. Grant, P.R. 1991. Natural selection and Darwin's finches. *Scientific American* 265 (October): 82–87.

Chasing *Tyrannosaurus* and *Deinonychus* Around the Tree of Life: Classifying Dinosaurs

Thomas R. Holtz, Jr.
Department of Geology
University of Maryland, College Park

We owe the word "dinosaur" and its formal version Dinosauria, meaning "fearfully great lizard," to Sir Richard Owen. In 1842 he published a paper that described the fossil reptiles of Great Britain, and he noted that three were sufficiently different from all other reptiles to warrant a particular group name of their own: carnivorous *Megalosaurus*, herbivorous *Iguanodon*, and armored *Hylaeosaurus*.[1] In so doing, Owen began the study of classifying dinosaurs.

Classification facilitates conversation. In order to talk about anything, we have to have names and labels: words that refer to items we are discussing. This is true whether we are talking about sports teams, flavors of ice cream, emotions, or dinosaurs. Furthermore, we group these items into larger categories according to different rationales: teams by sport, league, or hometown; flavors of ice cream into fruits and non-fruits; emotions into "bad" (such as anger and hatred) versus "good" (such as happiness and love); and so forth. In other words, we look for a taxonomy (a system of names) and a scheme of classification.

> *The methods of cladistic analysis continue to hold great promise for our understanding of the evolution of the different groups of dinosaurs.*
>
> *Tom Holtz*

Classification Schemes

Dinosaurs, as animals, are given names and are classified in the same way that all organisms are named and classified. Prior to the 1700s there was no single set of rules of taxonomy used by scientists. Instead, different cultures and different individuals in each culture organized animals, plants, fungi, and other organisms into particular schemes based on different attributes, such as the usefulness of the organism to humans, the danger it posed to humans, or its attractiveness. Carl von Linné, better known by the Latin form of his name, Linnaeus, developed the basic set of rules of biological nomenclature used by scientists since the mid-1700s. Linnaeus observed that the diversity of living things could be organized into a Natural System[2], based on the features

Thomas R. Holtz, Jr. is a dinosaur paleontologist in the Department of Geology at the University of Maryland, College Park. He received his Ph.D. in geology and geophysics at Yale University. His primary research interests are the evolution and adaptations of theropod dinosaurs, especially the Tyrannosauridae (tyrant dinosaurs); the ecomorphology of predation; and the effect of plate tectonics on Mesozoic terrestrial vertebrate distributions. He is Director of the College Park Scholars Earth, Life & Time program, a two-year honors program for undergraduates interested in natural history.

present or absent in the physical form and on the behavior of the organisms. Linnaeus' system was universal. It could be applied to all organisms—plants, animals, fungi, and even bacteria. It was also international. The system established a set of names in Latin or Greek, or rendered into a Latinate form, to be used by scientists, regardless of their native tongue.

Nested Hierarchies

The Linnaean system was organized as a nested hierarchy—a set of categories within larger categories within larger categories, like boxes within boxes within boxes. Each named group, or taxon (plural taxa), is a unit of biological diversity. Small units, like the species *Tyrannosaurus rex*, are grouped into larger units, such as the genus *Tyrannosaurus*, which are grouped into even larger units, for example Tyrannosauridae, Theropoda, Saurischia, and Dinosauria.

Linnaeus himself wrote, at least in his early works, that he considered species to be fixed; that is, species did not change into other species. To him, the nested hierarchy was a useful descriptor of the diversity of life because it reflected the organized mind of the Creator. A century later in his 1859 masterpiece *The Origin of Species*,[3] Charles Darwin recognized the underlying reason for this pattern—common descent with modification. Darwin discovered that organisms evolve in response to selection of some variations in a population relative to the other variations. His idea of evolution by Natural Selection is sometimes over-simplified as changes in a single lineage through time, but he also recognized it as responsible for larger patterns. Specifically, more than one set of variations in an ancestral population might preferentially survive relative to the rest of their kin. Over time different sets of variations (physical or behavioral attributes) would be selected for in the two different subpopulations. Eventually, these two subpopulations would be so different from each other that they would not be able to interbreed, and would therefore represent new species. Thus, Darwin recognized a mechanism by which a single common ancestral population could give rise to two or more new species, which themselves could sur-

vive and perhaps diverge into more species, or alternatively fail to survive and become extinct.

Darwin realized that this pattern of divergence of lineages through time was the reason that naturalists could uncover a nested hierarchical pattern to groups of organisms. For example, lions and tigers (*Panthera leo* and *Panthera tigris*, to give the formal names of these taxa) are more similar to each other than either is to grizzly bears (*Ursus arctos*), because lions and tigers share a more recent common ancestor with each other than they do with bears. Darwin recognized that the reason for the hierarchical nature of life was the tree-like structure of the history of life. He envisioned the stems of the tree representing common ancestors in the past whose descendants branched into different lineages, some going extinct, others surviving and perhaps developing additional branches; and he saw all evolving new and distinctive features in response to selection from the world around them. Closely related forms had diverged more recently from a common ancestral population, while distantly related forms represented branches that had split off further down the Tree of Life.

In *The Origin of Species* Darwin proposed that the Linnaean system would have to be modified to recognize that the organizing principle of taxonomy is ultimately "propinquity of descent"—that is, patterns of common ancestry. Darwin envisioned a method of classification where "more closely related to" meant "shared a more recent common ancestor with," rather than simply "is more similar in appearance to." Darwin recognized in *The Origin* that similarity might not always be a reliable indicator of shared ancestry. There are examples of anatomical or behavioral features evolving independently, also called convergently, in separate branches of the Tree of Life, particularly if those features were responses to similar environmental conditions. For example, animals that live in aquatic environments tend to evolve streamlined shapes, regardless of common ancestry. Therefore, he cautioned that new approaches to classification should take a look at many characters in combination, in the hope that the actual historical pattern of ancestry will show up.

The Method of Cladistics

After several attempts over the intervening decades to search for a new approach to classification that would reflect such patterns of ancestry, a German entomologist developed the organizing scheme used by most biologists today.[4,5] This entomologist, Willi Hennig, recognized that it would be impossible to know every single detail of the Tree of Life. Most individual animals and plants are eaten or decay before they can possibly be fossilized; indeed, many species of organisms will never be preserved in the fossil record because they did not live in environments where they would be likely to be buried, preserved, fossilized, and later discovered by paleontologists. However, Hennig recognized that the shape of the Tree of Life would reflect primitive features, unique features, shared derived features, and convergent features. Some features are present in all the taxa being studied. Fig. 1 shows that in a set of four carnivorous dinosaurs (*Allosaurus*, *Deinonychus*, *Albertosaurus*, and *Tyrannosaurus*) all had three features in common: (1) a hinge in the middle of the lower jaw, (2) a wishbone, and (3) bipedality. Since they shared these features, then presumably the same features would be found in the common ancestor of these four dinosaurs, and hence they are considered to be primitive (or ancestral) features. Alternatively, the most recent common ancestor of *Allosaurus*, *Deinonychus*, *Albertosaurus*, and *Tyrannosaurus* may have lacked a hinged jaw, lacked a wishbone, and walked on all fours, and in which case each of these four dinosaurs evolved

Fig. 1.

Four carnivorous dinosaurs, and observations of their features.

	Allosaurus	Deinonychus	Albertosaurus	Tyrannosaurus
1) Hinge in lower jaw	Yes	Yes	Yes	Yes
2) Wishbone	Yes	Yes	Yes	Yes
3) Bipedal	Yes	Yes	Yes	Yes
4) Retractable sickle claw	No	Yes	No	No
5) Backwards-pointing pubis	No	Yes	No	No
6) Number of fingers	3	3	2	2
7) Third metatarsal in foot	Unpinched	Unpinched	Pinched	Pinched
8) Astragalus (ankle bone)	Short	Tall	Tall	Tall
9) Tip of ischium	Expanded	Pointed	Pointed	Pointed

relative recency of common ancestry and this ancestry can be approximated by the distribution of features among the organisms that are available. Hennig's method became known as cladistics (from clade, or "branch"), because it is primarily concerned with recovering the branching order of common ancestry. The method of cladistics is the search for the simplest distribution of derived features to approximate the historical branching pattern of the Tree of Life.

Different Types of Features

Hennig noted that the features found in a set of organisms fall in four general categories:

the features independently. However, that would require more evolutionary changes than the more straightforward, simpler explanation that they had all three features in common. Primitive features reveal that the taxa are related at some level, but they don't help us resolve who is more closely related to whom.

Unique features are those found in only one of the taxa being examined. For example, among the four dinosaurs considered, two features are found only in *Deinonychus*: (4) retractable sickle claw on the foot and (5) a backward-pointing pubis (Fig. 1). These features must have evolved after the ancestor of

Deinonychus split from all the other meat-eating dinosaurs in this study. They may be helpful in recognizing *Deinonychus*, but they don't help us to determine which of the remaining three dinosaurs was the closest relative of *Deinonychus*. It should be noted that the use of "unique" or "primitive" for these features applies only to this particular analysis. For example, *Velociraptor* also had a retractable sickle claw and a backward-pointing pubis. Had it been included in the present study, these features would be shared between *Velociraptor* and *Deinonychus*, rather than limited to the latter.

The remaining features of the original set are found in more than one, but not all, of the dinosaurs being studied. These features evolved after the ancestors of the four dinosaurs in this study began to diverge from a common ancestor, and are called derived features. Hennig recognized that derived features potentially serve as clues to help discover the branching pattern of the Tree of Life, because they evolved on a lineage leading to some, but not all, of the taxa being studied. Hennig also recognized that there are two subsets of derived characters. Some features might be convergent. These convergent features would not help us discover patterns of common ancestry because they were evolved independently. The other features, though, may be shared derived features: evolutionary novelties inherited from a common ancestor. Hennig understood that by discovering the pattern of shared derived features, we might approximate the branching order of the lineages leading to the creatures we are studying.

Reconstructing the Tree of Life

Several features in this example represent derived, but not unique characters (Fig. 1). These include (6) a two-fingered (rather than three-fingered) hand, (7) a pinched third metatarsal in the foot, (8) a tall astragalus ankle bone, and (9) a pointed ischium. None of these evolutionary novelties occurs in *Allosaurus* (and indeed are lacking in meat-eaters more primitive than *Allosaurus*, such as *Torvosaurus*, *Ceratosaurus*, and *Coelophysis*). This observation suggests that the lineage leading to *Allosaurus* represents the first of all the forms in this study to branch off from the others. Thus, the features found in *Allosaurus* represent the ancestral condition for the four taxa in this analysis. However, it is important to understand that we do not regard *Allosaurus* as the actual ancestor of any of the dinosaurs in question!

Of the remaining three, *Deinonychus*, *Albertosaurus*, and *Tyrannosaurus*, it is likely that one pair of these dinosaurs had a more recent common ancestor than either did with the third. There are in fact three possibilities for this situation:

1) *Deinonychus* and *Tyrannosaurus* shared a more recent common ancestor with each other than with *Albertosaurus* (Fig. 2A);

2) *Deinonychus* and *Albertosaurus* shared a more recent common ancestor with each other than with *Tyrannosaurus* (Fig. 2B); or

3) *Tyrannosaurus* and *Albertosaurus* shared a more recent common ancestor with each other than with *Deinonychus* (Fig. 2C)

We do not know ahead of time which of these three possibilities is correct, but we can construct our hypothesis based upon the evidence. The criterion for determining which pairing is most likely is simplicity. All other things being equal, we choose the simplest answer. This criterion, known more formally as parsimony, is a standard principle in science. When multiple possible explanations exist, the one that requires the fewest assumptions is likely to be more nearly correct.

The three possibilities can be represented as branching diagrams called cladograms (Fig. 2). A cladogram is the graphic representation of a hypothesis of the relationships between taxa. Unlike traditional evolutionary family tree drawings, in which the vertical lines represented the actual group of reproducing organisms through time, the lines (branches) and the nodes joining the branches are just place holders to represent common ancestry. For example, in Fig. 2C, there is a single node marked "a" that represents the shared ancestry of *Albertosaurus* and *Tyrannosaurus*. The next node towards the base of this tree, marked "b," represents that there is a common ancestor

shared by *Deinonychus* and the lineage leading to *Alberto-saurus* and *Tyrannosaurus*. And finally, there is a node "c" that represents a common ancestor shared by *Allosaurus* and all remaining forms.

Choosing the Best Hypothesis

As explained, only shared de-rived features can actually in-form us about the pattern of shared ancestral relationships. For any given set of taxa, more than one solution is possible (that is, more than one possible cladogram), each representing a different hy-pothesized relationship. By counting the number of evolu-tionary changes required to produce each of these trees, we can find the most parsimoni-ous (simplest) solution to the distribution of features we see.

Using our example in Fig. 2A, we find that this clado-gram is explained by 11 evo-lutionary changes: one change each for the three primitive features shared by all; one change each for the two unique *Deinonychus* features; and the remaining two features each evolving independently—once each on the line leading to *Tyrannosaurus*, and once each on the line leading to *Albertosaurus*. Similarly, the cla-dogram in Fig. 2B shows a similar distribution requiring 11 changes, but in this case the changes in features 8 and 9 are explained as re-versals (evolving back from the derived state), having been present in the common ancestor of *Tyrannosaurus*, *Albertosaurus*, and *Deinon-ychus*, then reversed along the line leading just to *Deinonychus*. Finally, the cladogram in Fig. 2C requires only 9 steps, with features 6 and 7 being present in the common ancestor of *Deinonychus*, *Albertosaurus*, and *Tyrannosaurus*, and features 8 and 9 evolving only a single time—after the common ancestor of *Alberto-*

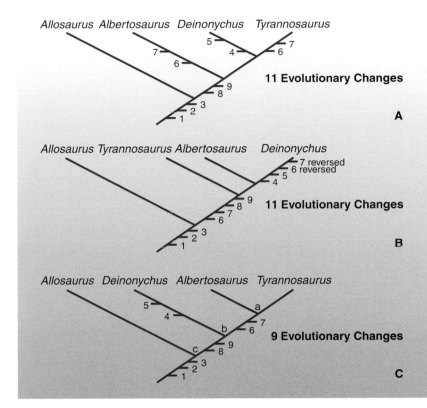

Fig. 2. The three possible different cladograms for the dinosaurs from Fig. 1, showing the distribu-tion of derived features in each. Numbers along the branches represent the various features from Fig. 1. Note that the cladogram in **2C** requires the fewest number of evolutionary changes, and so would be the one preferred in this analysis.

saurus and *Tyrannosaurus* split off from the line leading to *Deinonychus*, but before the ances-tors of *Albertosaurus* and *Tyrannosaurus* split from each other. So, all other things being equal, we would choose the cladogram in Fig. 2C, with the fewest number of evolutionary changes, as representing the closest approxima-tion to the actual historical pattern of common ancestry.

Is the cladogram in Fig. 2C the true histori-cal pattern? We can't know for certain. It is only our best hypothesis for the pattern, given the data at hand. An advantage to the cladistic method is that its results can be subjected to further tests. For example, new taxa or new features can be added to the analysis, and if the simplest distribu-tion of derived features in the new analysis matches the previous results, the original hypoth-esis is supported. If not, the original hypothesis would be rejected in light of new observations.

Advantages of the Cladistic Method

That new analyses can be run with the addition of new data gives cladistic analyses distinct advantages over the traditional method of simply connecting possible ancestors and possible descendants through time. A cladistic analysis searches for patterns of relative common ancestry, something we can be more secure about. At some level there is a common ancestor for every pair of species that ever existed. Neither *Triceratops*, nor a dog, nor a dandelion is an ancestor or descendant of the others, but we can still recognize that the first two shared a common ancestor not shared by the plant. Thus, even if some, or even many, of the individual species along the actual branches of the Tree of Life are missing, we can approximate its shape, understanding that additional data may change that approximation.

Additionally, the nature of cladistic analyses makes them relatively easy to conduct using computer programs. After a scientist has coded the observations of derived features for various taxa and entered those data into a search program, a computer can sort through millions, or even billions of trees, searching for the best—most parsimonious—results. So, more and more complex analyses, each with dozens of taxa and hundreds of features, can be conducted far faster than a human being could ever hope to achieve with a pad of paper and a pencil. Such studies are now commonplace throughout biology and paleontology, using not only the shapes of bones, but also soft tissues, DNA sequences, and even behavioral features. These studies often result in many equally simple cladograms, all of which are equally good approximations given the data available. Better approximations will be reached by running analyses as more evidence is gathered.

Cladistic Classifications

Darwin put forth the idea that a good organizing principle for classification would be patterns of common ancestry. Hennig's method of cladistic analysis allows the most likely pattern of common ancestry to be recovered and used as a basis of classification.[6]

One method, favored by many biologists (including the majority of dinosaur paleontologists), is a scheme proposed by Hennig himself, in which all taxon names represent clades (complete branches of the Tree of Life). Each named group represents an ancestor and all of its descendants. For example, on cladogram Fig. 2C, node "a" would represent the clade Tyrannosauridae containing, in this very reduced cladogram, only *Tyrannosaurus* and *Albertosaurus*. Node "b" would be the larger group Coelurosauria, containing Tyrannosauridae and *Deinonychus*; and finally node "c" would be Avetheropoda (*Allosaurus* plus the coelurosaurs). Over time, some members within a clade might become more and more transformed from the ancestral condition as new features evolved. However, they would still be considered part of the clade because they descended from that common ancestor.

So, What Are Dinosaurs? A Basic Classification

Using the methods outlined above, the position of dinosaurs within the Tree of Life (Fig. 3) and the details of the cladogram of the dinosaurs themselves (Fig. 4) have been reconstructed. Although broad consensus exists about most of the details, there are sometimes discrepancies in the exact results obtained in different studies.[7, 8, 9, 10] These discrepancies may reflect the choice of features selected for comparison, or may simply be a reflection of lack of information about some aspects of the fossils in question.

Cladistic analyses of the tetrapods consistently show that dinosaurs are a subgroup within the archosaurian reptiles, a group that includes crocodilians among living forms, as well as many extinct creatures.[11] Note that calling dinosaurs "reptiles" in the cladistic sense tells us nothing about whether dinosaurs were cold-blooded or warm-blooded. Like all cladistic names, Reptilia is a group defined by common ancestry, rather than by a general grade of organization. Dinosaurs and their closest relatives, such as *Marasuchus*, are distinguished from all other reptiles by a fully upright stance of the hindlimbs and a simple ankle joint. Fur-

thermore, Dinosauria itself is distinguished by an open hip socket. Because of the incompleteness of the fossils of the immediate closest relatives of dinosaurs, there is some uncertainty as to what other features characterize the common ancestor of all dinosaurs.

Two main branches exist within Dinosauria: Ornithischia and Saurischia. The ornithischians include *Iguanodon* and *Hylaeosaurus* among the original three members of Owen's Dinosauria, as well as all the dinosaurs sharing a

the plated Stegosauria, tank-like Ankylosauria, and their closest relatives; the ridge-headed Marginocephalia, including the dome-skulled Pachycephalosauria and the parrot-beaked Ceratopsia (most famously including *Triceratops* and the other horned

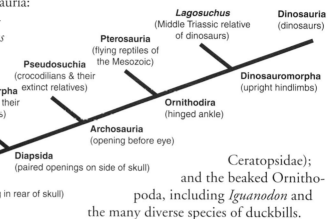

Fig. 3. A cladogram showing the position of dinosaurs among the four-limbed vertebrates.

Ceratopsidae); and the beaked Ornithopoda, including *Iguanodon* and the many diverse species of duckbills.

The Saurischia include *Megalosaurus* and all the dinosaurs more closely related to it than to *Iguanodon*. One clade of saurischians is the long-necked herbivorous group Sauropodomorpha. Sauropodomorphs include primitive bipedal forms like *Saturnalia* and *Plateosaurus* as well as the gigantic quadrupedal Sauropoda (such as *Apatosaurus*, *Brachiosaurus*, and *Argentinosaurus*). True sauropods are the largest animals ever to walk the land.

Perhaps the most diverse clade of dinosaurs is the Theropoda. Theropods contain the various carnivorous dinosaurs, from tiny *Compso-*

more recent common ancestor with them than with *Megalosaurus*. At present all known ornithischians seem to have been herbivorous. In fact, they all had an extra jawbone (the predentary) which held part of a horny beak for chopping up plants. All but the most primitive had a backward-pointing pubic bone, presumably allowing for the increased gut space necessary to digest large amounts of vegetation. Major groups of ornithischians include the armored Thyreophora, comprising

Fig. 4. A basic cladogram of the relationships among the major groups of dinosaurs.

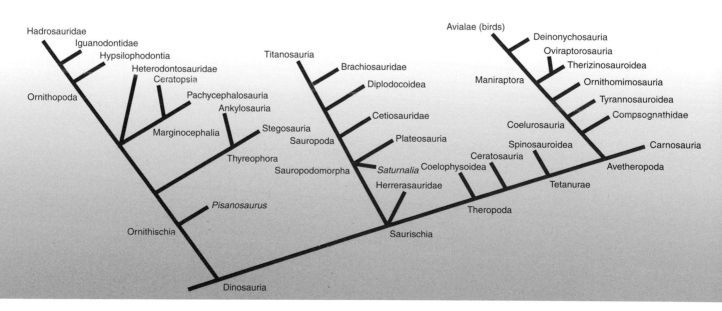

gnathus and *Microraptor* through giants such as *Tyrannosaurus*, *Spinosaurus*, and *Giganotosaurus*. Birds represent descendants of small carnivorous dinosaurs (see Currie, page 89). Indeed, by the principles of cladistic classification birds are members of the Dinosauria.

The methods of cladistic analysis continue to hold great promise for our understanding of the evolution of the different groups of dinosaurs. New studies continue to determine more precisely the relationships of the various species of dinosaurs in each of the major groups. Furthermore, by using the principles of parsimony and an understanding of the shape of the Tree of Life, paleontologists are attempting to discover the pattern of evolution of various aspects of the biology and behavior of dinosaurs. Recovering the interrelationships among the dinosaurs and their position among the vertebrates is only the first step in exploring the evolution of this amazing group of reptiles.

References

1. Owen, R. 1842. Report on British fossil reptiles, Part II, 60–204. *Report of the Eleventh Meeting of the British Association for the Advancement of Science for 1841.*

2. Linné, C. 1758. *Systema natura per regina tria naturae, secundum classes, ordines, genera, species cum characterisbus, differentiis, synonymis, locis.* Editio decima, reformata, Tomus I: Regnum Animalia. Stockholm: Laurentii Salvii.

3. Darwin, C. 1859. *The origin of species by means of natural selection, or the preservation of favoured races in the struggle for life.* London: John Murray. (Complete text of the first edition is available online at www.literature.org/authors/darwin-charles/the-origin-of-species/)

4. Hennig, W. 1950. *Gründzuge einer theorie der phylogenetischen systematik.* Berlin: Deutscher Zentralverlag.

5. Hennig, W. 1966. *Phylogenetic systematics.* Urbana: University of Illinois Press.

6. Holtz, T.R., Jr., and M.K. Brett-Surman. 1997. The taxonomy and systematics of the dinosaurs. In *The complete dinosaur,* eds. J.O. Farlow and M.K. Brett-Surman, 92–106. Bloomington: Indiana University Press.

7. Sereno, P.C. 1997. The origin and evolution of dinosaurs. *Annual Review of Earth and Planetary Sciences* 25:435–489.

8. Sereno, P.C. 1999. The evolution of dinosaurs. *Science* 284:2137–2147.

9. Holtz, T.R., Jr. 2000. Classification and evolution of the dinosaur groups. In *The Scientific American book of dinosaurs,* ed. G.S. Paul, 140–169. New York: St. Martin's Press.

10. Pisani, D., A.M. Yates, M. Langer, and M.J. Benton. 2002. A genus-level supertree of the Dinosauria. *Proceedings of the Royal Society Series B* 269:915–921.

11. Brochu, C. 2001. Progress and future directions in archosaur phylogenetics. *Journal of Paleontology* 75:1185–1201.

It Takes Time: Dinosaurs and the Geologic Timescale

Spencer G. Lucas

New Mexico Museum of Natural History
Albuquerque

*S*tegosaurus lived 150 million years ago. Dinosaurs became extinct at the end of the Cretaceous Period. What do these statements mean? How do we know how long ago any dinosaur lived? This chapter answers these questions by explaining how geologists measure geologic time and where dinosaurs fit in the geologic timescale.[1]

The geologic timescale (Fig. 1) divides the last 4.6 billion years of Earth history into named intervals while also providing numerical estimates (usually in millions of years) of their duration. Thus, it is actually two timescales in one—a relative scale of time intervals arranged chronologically and hierarchically (shorter intervals are grouped into longer ones), and a numerical timescale calibrated in millions of years.

Relative Timescale

About 1800, William Smith, a British civil engineer engaged in canal building, realized that a given layer of sedimentary rock usually contains distinctive fossils. Smith traced many of these distinctive fossils over large areas, even when the nature of the enclosing rocks changed, and this allowed him to construct the first geological map of England.[2]

Parallel to Smith's work, French comparative anatomist Georges Cuvier and geologist Alexander Brongniart independently discovered that distinctive kinds of fossils are commonly associated with specific rock layers. Whereas Smith did not attach geologic time significance to his observations, Cuvier, Brongniart, and other French paleontologists of the early 19th century did. They concluded that each rock layer with its distinctive fossils, and its position relative to other rock layers, represents a particular "stage" in the history of life. This conclusion formed the basis for using fossils to identify intervals of geologic time.[3]

In the ensuing 200 years the relative geological timescale developed as geologists and paleontologists divided geologic time into intervals that can be

> *. . . radioactivity, the spontaneous decay (falling apart) of some types of atoms, provides a natural clock for estimating numerical geologic ages, because the rate of decay is constant for a given type of atom.*
>
> Spencer Lucas

Spencer G. Lucas is Curator of Paleontology and Geology at the New Mexico Museum of Natural History in Albuquerque. He is also Adjunct Professor of Geology at the University of New Mexico. He received a B.A. from the University of New Mexico and a M.S. and Ph.D. in geology from Yale University. His research focuses on the use of fossil vertebrates in biostratigraphy and correlation. He is the author of *Dinosaurs: The Textbook,* published by McGraw-Hill Company.

Fig. 1. Geologic Time Scale (based on the official timescale of the International Union of Geological Sciences)

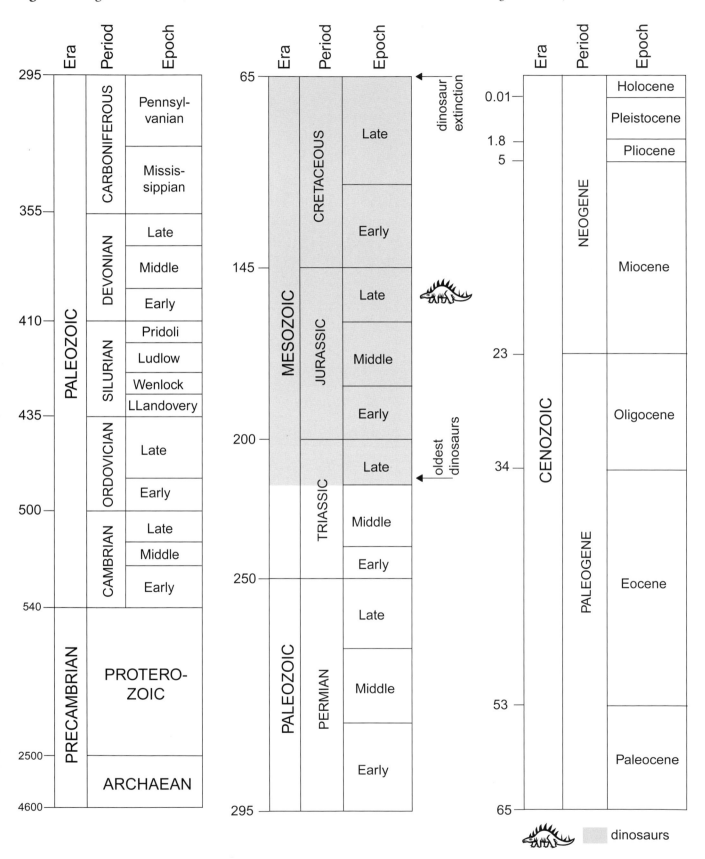

recognized globally.[4] The resulting relative geologic timescale (Fig. 1) is a hierarchy of time intervals, from long (eon, era) to short (period, epoch, age). The name of each interval has historical significance, reflecting a concept held by the geologist who coined the term. For example, the eon name "Phanerozoic" refers to the interval of Earth history when life was very evident (Greek *phaneros,* evident) on the planet, as proven by an abundance of fossils. During the preceding era, the Proterozoic, life was in one of its earliest stages of evolution (Greek *proteros,* earlier). In 1841, British geologist John Phillips coined the names of the Phanerozoic eras—Paleozoic, Mesozoic, and Cenozoic—as time intervals of ancient life (Greek *palaios,* ancient; *zöe,* life), intermediate life (Greek *mesos,* intermediate) and recent life (Greek *kainos,* recent). The Paleozoic-Mesozoic and the Mesozoic-Cenozoic boundaries were defined by major, global extinctions inferred from the fossil record.

During the 19th century, European geologists named the geologic time periods.[2] Some refer to a place (e.g., Permian, based on Perm in Russia; Devonian, for Devon in England), some to a kind of rock (e.g., Carboniferous, for coal; Cretaceous, for Latin *creta,* chalk), and others are more colorful references (Ordovician and Silurian for ancient Welsh tribes who fought the Romans). Tertiary (third) and Quaternary (fourth) are holdovers from an older geological timescale of the 18th century, when Mesozoic rocks were called "Secondary," and pre-Mesozoic rocks were "Primary." Currently, many geologists replace Tertiary and Quaternary with a more balanced division of Cenozoic time into Paleogene and Neogene.

Each geologic period is divided into shorter geologic time intervals, the epochs and ages. Of these, the names of the Cenozoic epochs (e.g., Paleocene, Eocene, etc.) are used most often. British geologist Charles Lyell originally coined some of these names in 1833, basing them on the fact that species of fossil molluscs (clams and snails) look more and more like modern species in progressively younger Cenozoic rocks.[2] In the oldest Cenozoic rocks, the fossil molluscs little resemble modern species, but by late Cenozoic time they

appear quite modern. So, Lyell aptly named the earliest Cenozoic the Eocene, "dawn of the recent" (Greek *eos,* dawn; *kainos,* recent). By Miocene time, though, a few molluscs resembled modern forms (Greek *meion,* less), and by the Pliocene there were many more such relatively modern molluscs (Greek *pleion,* more). The other epoch names were added later, as the Cenozoic timescale was refined.

Correlation

Regardless of the term applied to a time interval and the term's derivation, the definition and characterization of the time interval is universal. Thus, let's say a geologist identifies a body of rock in a particular place and asserts that the body of rock was formed during a particular time interval that has never before been defined. That place is referred to as the "type locality." The fossils that the rock contains are of plants and/or animals that lived during that time interval. To assign rocks (and fossils) or events from another place to that time interval requires correlation—establishing the equivalence of age of two rock bodies in separate areas. Geologists do this by demonstrating that the rock bodies are continuous, are very similar to each other, and/or contain the same fossils.

To correlate is to establish the age equivalence of geological events or rocks in the histories of separate regions.[3] Fossils valuable in correlation are called index fossils because they identify and determine the age of the rocks in which they are found. A good index fossil has a short geologic time range, thus, it is characteristic of a short interval of geologic time. At the same time, it is geographically widespread, common and easy to identify.

Defining the Boundaries

In the last 50 years, the focus of timescale refinement has shifted from defining the intervals to defining precisely the boundaries between time intervals. The goal is to define the beginning of a time interval; by default, that defines the end of the preceding interval. To assign a rock, fossil, or event to a given time interval, we need only demonstrate that it is older than the end of and younger than the beginning

of that interval.[5] For example, to assign a *Stegosaurus* fossil a Jurassic age, we must demonstrate that it is older than the end of the Jurassic and younger than the beginning of the Jurassic.

The beginning of a major time interval (eon, era, period) usually is associated with dramatic physical and/or biological changes on Earth. We often recognize these changes by shifts in the flora and fauna as shown in the fossil record. Thus, the beginning of the Mesozoic is marked by the greatest extinction of life in Earth history, and its end (i.e., the beginning of the Cenozoic) corresponds to the extinction of the dinosaurs and many other life forms. Such large-scale events usually provide widespread and easily recognized criteria for a geological time boundary.

The boundaries of shorter geological time intervals (epochs, ages) usually are not marked by such dramatic events. A single evolutionary event, such as the appearance of one or more new species, often identifies the boundary of a shorter time interval. For example, the appearance of a single species of ammonoid cephalopod (extinct relatives of squid and octopuses) is the principal criterion for identifying the boundary between the Middle and the Late Triassic.

Numerical Timescale

The numerical timescale assigns "absolute" ages to rocks in thousands, millions, or billions of years, based on radio-isotopic dating.[6, 3] Such numerical ages can be determined only on rocks that contain sufficient quantities of radioactive elements, usually some form of volcanic rock (ash or lava). Thus it is not possible to determine numerical ages for all rocks. It is also important to remember that the final numbers are approximations based on the best-calculated ages, and that all of these calculated ages have a margin of error. For this reason, as better estimates become available, ages change. For example, since 1980 the age estimate of the Permian-Triassic boundary has changed from 237 to 250 million years ago.

Before the discovery of radioactivity in the late 1800s, calculating accurate numerical ages for events in geologic history was impossible.

But, radioactivity, the spontaneous decay (falling apart) of some types of atoms, provides a natural clock for estimating numerical geologic ages, because the rate of decay is constant for a given type of atom. So, provided that the rate of decay is known for the radioactive atoms and that the rock contains a known quantity of such atoms, geologists can use this "radioactive clock" to determine the numerical ages of rocks.[6]

In laboratories, geochemists have precisely calculated the rates at which radioactive atoms decay. Many of these rates are very slow, so great laboratory precision is needed to calculate them. For example, the rate of decay of uranium-238 is so slow that it takes 4.5 billion years for half the uranium atoms in a sample of uranium to decay. Measuring the amount of decay of uranium atoms in a rock from the age of dinosaurs involves measuring a very small amount of material, so great care must be taken when those measurements are made. A given sample can commonly be age-dated using different radioactive elements, thus providing multiple independent tests of that sample. Comparing results of these tests gives us a higher level of confidence in its age.

What poses the largest problem for a numerical time scale is the fact that most rocks, indeed nearly all sedimentary rocks, do not contain enough radioactive atoms that are decaying for analysis. Sufficient quantities of these types of atoms are found almost exclusively in igneous rocks. However, igneous rocks do not usually contain fossils, so the only way to calculate a numerical age for a fossil is to find an igneous rock close to the sedimentary rock layer that contains the fossil. Ideally, sheets of lava above and below a layer of sedimentary rock containing dinosaur fossils will yield numerical ages that bracket the age of the fossils. A *Stegosaurus* fossil found in a layer of rock between two lava flows—the lower one dated at 151 mya and the upper dated at 149 mya—may reasonably be said to be about 150 million years old. But, this does not happen often, and paleontologists can only estimate the numerical age of a dinosaur fossil by evaluating its proximity to the nearest rocks that yield a numerical age or to the nearest layer containing an index fossil of known age.

Such evaluation relies on several lines of evidence used in correlation. But, the fact remains that we cannot assign precise numerical ages to all dinosaur fossils, and we may never be able to do so. Therefore we continue to use the divisions of the relative geologic time scale, combined with the best numerical-age estimates available, when discussing the "when" and the "how fast" of the age of dinosaurs.

Dinosaurs in the Geologic Timescale

Dinosaurs lived during the Mesozoic, so we place dinosaurs in the geologic timescale in the three geologic periods of the Mesozoic: the Triassic, Jurassic, and Cretaceous (Fig. 1).[1]

German geologist Friedrich August von Alberti coined the term Triassic in 1834. In studying the salt deposits of Germany, Alberti found three different rock sequences, an older one mostly of sandstone, an intermediate one mostly of limestone, and a younger one mostly of shale. All three sequences were younger than rocks identified as Permian, but older than Jurassic rocks. Thus, Alberti established a distinct time interval between the Permian and Jurassic, naming it Triassic (*triad* is Latin for "three") for the three rock sequences in Germany. The Triassic, of course, is subdivided into three time intervals, the Early, Middle, and Late Triassic epochs. Today, rocks of Triassic age are recognized worldwide, and the Triassic Period is estimated to have lasted about 50 million years, from about 250 to 200 million years ago (Fig. 1). The oldest dinosaur fossils are of Late Triassic age, about 225 million years old.[1] Dinosaurs appeared then, at almost exactly the same time as the first turtles, crocodiles, pterosaurs (flying reptiles), plesiosaurs (long-necked marine reptiles), and mammals. The meat-eater *Coelophysis* and the plant-eater *Plateosaurus* are examples of Triassic dinosaurs.

Alexander von Humboldt was a famous explorer and geographer of the late 18th century. He was also a trained geologist who, in 1799, first used the name Jura for a distinctive limestone in the Jura Mountains of Switzerland. This became the basis for the word Jurassic, used by geologists to refer to the time period between the Triassic and the Cretaceous, 200 to 145 million years ago (Fig. 1). Like the Triassic, the Jurassic is divided into Early, Middle, and Late intervals. Dinosaurs flourished everywhere on Earth during the Jurassic and were both abundant and diverse. Indeed, the largest of all dinosaurs lived during the Jurassic. Some of the most famous dinosaurs are of Jurassic age, including *Allosaurus*, *Brachiosaurus*, "*Brontosaurus*" (*Apatosaurus*) and *Diplodocus*. *Stegosaurus* fossils are found in 150-million-year-old rocks of Late Jurassic age, so we can say that *Stegosaurus* lived about 150 million years ago.

In parts of western Europe, especially in Great Britain, France, and Belgium, rocks that are younger than Jurassic and older than Cenozoic are mostly chalk. For this reason, a Belgian geologist, J.J. D'Omalius d'Halloy, used the French term *Terrain Cretacé* (Cretaceous System) in 1822 to refer to these rocks (*creta* is Latin for "chalk"). The Cretaceous Period is the interval of geologic time between the Jurassic and Cenozoic, 145 to 65 million years ago (Fig. 1). It has traditionally been divided into Early and Late time intervals, but, during the last 20 years, many geologists have also distinguished a Middle Cretaceous time interval. Dinosaurs are known from rocks of Cretaceous age on all the world's continents. Well-known Cretaceous dinosaurs include *Ankylosaurus*, *Hadrosaurus*, *Triceratops* and *Tyrannosaurus*. These dinosaurs became extinct at the end of the Cretaceous, about 65 million years ago.

For online instruction in the methods of numerical age estimates of rocks, go to:
- vcourseware5.calstatela.edu/VirtualDating/index.html
- www.talkorigins.org/faqs/isochron-dating.html

R e f e r e n c e s

1. Lucas, S.G. 1991. Dinosaurs and Mesozoic biochronology. *Modern Geology* 16:127–138.

2. Berry, W.B.N. 1987. *Growth of a prehistoric time scale*, Revised edition. Palo Alto: Blackwell.

3. Prothero, D.R. 1990. *Interpreting the stratigraphic record*. New York: W.H. Freeman & Company.

4. Harland, W.B., R.L. Armstrong, A.V. Cox, L.E. Craig, A.G. Smith, and D.G. Smith. 1990. *A geologic time scale 1989*. Cambridge: Cambridge University Press.

5. Salvador, A. (ed.). 1994. *International stratigraphic guide*, Second edition. Boulder: International Union of Geological Sciences and Geological Society of America.

6. Dalrymple, G.B. 1991. *The age of the Earth*. Stanford: Stanford University Press.

Where Dinosaurs Roamed the Earth

Catherine A. Forster
Department of Anatomical Sciences
Stony Brook University, Stony Brook, NY

6

Catherine Forster
received her Ph.D. in Geology at the University of Pennsylvania, and worked at the Field Museum of Natural History and the University of Chicago before taking her present position as a professor at Stony Brook University. Her research interests center around dinosaur systematics, but they include research into other Mesozoic vertebrates as well as biogeography. She is actively involved in fieldwork in Madagascar and western China. In her spare time, Catherine enjoys gardening, home renovation, and pottery.

Dinosaurs arrived on the scene some time in the Late Triassic Period, about 230 million years ago. The first dinosaur originated at a single spot on Earth (although we don't know exactly where) as a minor member of a fauna dominated by other animals. By the beginning of the Jurassic Period, approximately 30 million years later, the dinosaurs had evolved into a rich array of forms and dominated faunas worldwide. They had also become a global phenomenon, having spread from their point of origination to every part of the planet. They continued to rule the global scene for another 140 million years, ever evolving and spreading over the Earth.

Dinosaurs lived just about everywhere during the Mesozoic Era. Their remains have been found on every continent, from the Antarctic Peninsula and north to well beyond the Arctic Circle. Not only were dinosaurs geographically widespread, they also conquered a wide array of environments, from hot, dry deserts to dense forests. We know where dinosaurs once lived from the evidence they left behind—fossilized skin and skeletal remains, footprints, eggs, and even excrement (coprolites).

We can study the distribution of dinosaurs in two basic ways: by looking at their precise geographic locations (their location in "space") at each specific point in time, and by looking at how these distributions change through time. The study of the distribution of animals through space and time is called paleobiogeography (literally: the geography of ancient organisms). This chapter presents an overview of dinosaur paleobiogeography: where they were found and when. With these data in hand, we can begin to ask additional questions about their lives, such as: what do these distributions tell us about the movements and evolutionary history of dinosaurs? Do the locations of specific dinosaurs tell us anything about their lifestyles? How were the distributions of dinosaurs affected by the evolution of the Earth and its environments and climates?

> *We need to be aware that as the dinosaurs evolved and moved around through the Mesozoic, the Earth was undergoing its own revolution in shape and configuration.*
>
> Cathy Forster

Where and When

The first step in understanding dinosaur distributions is to see exactly where their remains have been found. Commonly, these remains consist of partial to complete fossilized bones and skeletons. In some cases, other tantalizing remains of dinosaurs can be found, including eggs and eggshell fragments, skin impressions, footprints and trackways, and even coprolites. These remains must be documented for:

■ their precise geographic location;

■ the age of the rock in which they lie; and

■ which dinosaur species the remains represent.

This provides paleontologists with three crucial data points: the kind of dinosaur, where it once lived, and when it lived there.

Today, locating a dinosaur fossil in space can be accomplished with a global positioning system, which provides a precise latitude and longitude for any spot in the world. Since the age of the dinosaur is determined by the age of the rock, careful dating of the rock unit must occur. Although many rock units can be precisely dated by using radiometric methods, the ages of some rock units remain poorly known, and thus the ages of some dinosaurs can only be estimated.

Determining the species of dinosaur that belong to the remains is usually straightforward, especially when skeletal material is found. Paleontologists compare new skeletal remains to known specimens to determine whether they match established dinosaur species. If they don't match, the remains are named as a new species. Although evaluating skeletal materials is fairly straightforward, connecting coprolites, footprints, and eggshells to a specific dinosaur is often impossible. For example, features of a footprint may allow it to be classified as that of a theropod dinosaur, but may not let us know exactly which species of meat-eater made it.

Table 1 provides an inventory of areas around the world that have produced dinosaur remains. This distribution list is divided into time periods (Epochs), ranging from (oldest to youngest) the Late Triassic through the Late Cretaceous Periods (Fig. 1). For Canada and the United States, this list is further broken down into provinces and states. This list is general using broad time periods and large geographic areas—much more precise geographic and temporal data are generally used by paleontologists in their studies. For example, time periods are broken down into shorter segments, and geographic areas are subdivided into smaller parcels. The use of more restricted temporal and spatial units allows paleontologists to examine their data with added precision.

Remember that each country occurrence in Table 1 may represent anything from one specimen and locality, to scores of specimens and a host of localities. This is especially true for some countries, such as Russia, Australia, and China, which, like the United States and Canada, cover vast territories. For example, in the Late Cretaceous, both South Africa and Argentina are listed as producing dinosaur fossils. The South African datum consists of a single discovery of two tail vertebrae from a sauropod dinosaur. However, the Argentina datum includes hundreds of fossils discovered in dozens of localities throughout that country.

Although this current list provides the general "where and when" of dinosaur distributions, what is missing is the "who." Exactly which dinosaurs are represented by these occurrences is beyond the scope of this chapter (see Weishampel, 1990, for such a list)[1], but is nevertheless important for answering many paleobiogeographic questions.

Biases and Gaps: Are We Missing Some Dinosaurs?

Table 1 shows that we know a lot about where dinosaurs once lived, although we certainly don't know everything. What about the places and time periods where dinosaurs have not been found? Are these absences real, or are there other reasons why we may be missing part of the dinosaur fossil record?

We must remember that the dinosaur remains we find today probably represent only a small fraction of all the dinosaurs that once lived. We know only a small portion of the ages and places where dinosaurs once lived—so dinosaur occurrences worldwide tend to be

Table 1. Global Dinosaur Distribution

Late Triassic (235-208 mya)	North America	Canada (Northwest Territories, Nova Scotia); United States (Arizona, Colorado, Massachusetts, New Jersey, New Mexico, New York, North Carolina,Pennsylvania, Utah)
	Europe	Belgium, England, France, Germany, Italy, Scotland, Switzerland, Wales
	Greater Asia	China, India
	South America	Brazil, Argentina
	Africa	Lesotho, Madagascar, Morocco, South Africa
	Australiasia	Australia
Early Jurassic (208 -178 mya)	North America	Canada (Nova Scotia); United States (Arizona, Colorado, Connecticut, Massachusetts, New Jersey, Utah, Wyoming); Mexico
	Europe	England, France, Germany, Hungary, Portugal, Sweden
	Greater Asia	China, India, Iran
	South America	Brazil, Venezuela
	Africa	Algeria, Lesotho Morocco, Namibia, South Africa, Zimbabwe
	Antarctica	trans-Antarctic Mountains
Middle Jurassic (178-157 mya)	Europe	England, France, Portugal, Scotland
	Greater Asia	China
	South America	Argentina, Chile
	Africa	Morocco, Algeria
	Australasia	Australia
Late Jurassic (157-145 mya)	North America	United States (Alaska, Colorado, Montana, Oklahoma, New Mexico, South Dakota, Texas, Utah, Wyoming)
	Europe	England, France, Germany, Portugal, Spain, Switzerland
	Greater Asia	China, India, Thailand
	South America	Argentina, Chile, Columbia
	Africa	Madagascar, Malawi, Morocco, Niger, South Africa, Tanzania, Zimbabwe
	Australasia	New Zealand
Early Cretaceous (145-97 mya)	North America	Canada (British Colombia); United States (Alaska, Arkansas, Arizona, Colorado, Idaho, Kansas, Maryland, Montana, Nebraska, New Mexico, Oklahoma, South Dakota, Texas, Utah, Wyoming)
	Europe	Belgium, Croatia, England, France, Germany, Italy, Norway, Portugal, Romania, Spain
	Greater Asia	China, Georgia, Japan, Kazakhstan, Mongolia, South Korea, Thailand
	South America	Argentina, Brazil, Chile
	Africa	Algeria, Cameroon, Libya, Malawi, Mali, Morocco, Mozambique, Niger, Sudan, South Africa, Tunisia, Zambia, Zimbabwe
	Australiasia	Australia
Late Cretaceous (97-65 mya)	North America	Canada (Alberta, British Colombia, Northwest Territories, Saskatchewan, Yukon Territory); Honduras, Mexico, United States (Alabama, Alaska, Arizona, California, Colorado, Delaware, Georgia, Kansas, Maryland, Mississippi, Missouri, Montana, Nevada, New Jersey, New Mexico, North Carolina, North Dakota, Oregon, South Dakota, Texas, Utah, Wyoming)
	Europe	Austria, Belgium, Czech Republic, England, France, Netherlands, Portugal, Romania, Russia, Spain, Ukraine
	Greater Asia	China, India, Israel, Japan, Kazakhstan, Laos, Mongolia, Russia, Syria, Tajikistan, Uzbekistan
	South America	Argentina, Bolivia, Brazil, Chile, Colombia, Peru, Uruguay
	Africa	Algeria, Egypt, Kenya, Madagascar, Morocco, Niger, South Africa
	Australasia	New Zealand
	Antarctica	Antarctic peninsula

spotty and intermittent. For example, dinosaurs are still completely unknown in some areas of the world, while in others, such as China, they have been found in every time period from 235-63 mya (Fig. 1). There are a number of reasons for these "gaps" in our knowledge of dinosaur distributions:

1) Many places in the world have yet to be explored by paleontologists. This bias in fossil collection is due to a number of reasons, including inaccessibility of an area, inhospitable working conditions (disease, heat, wind, terrain, etc.), and unfriendly political climates.

Doubtless, new dinosaur fossils are awaiting discovery all around the world. As we continue to explore new areas and new rock layers, more dinosaurs are sure to come to light. For example, a group of paleontologists working in southern Madagascar recently discovered dinosaurs in Late Triassic rocks—the first known from that time period in that country.[2] In many cases, the dinosaurs are there, just waiting for a paleontologist to come along and find them.

It is not surprising that many of the best-explored areas occur in Europe and North

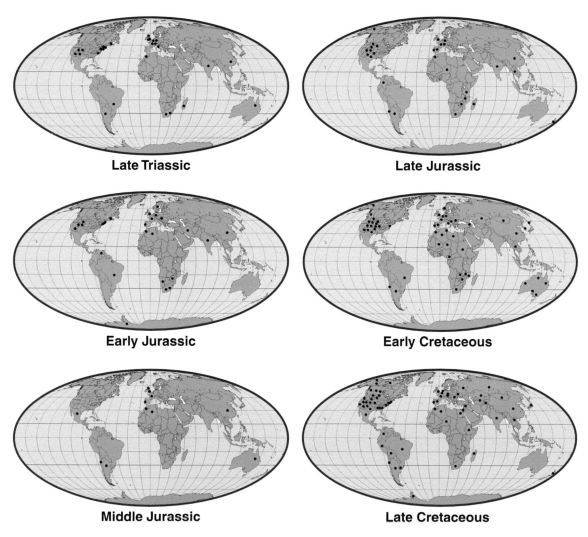

Fig. 1. These six maps show the configuration of the present day continents through the Age of Dinosaurs. Known dinosaur distributions per country are indicated with a black dot. The general time periods depicted are Late Triassic (Norian, 220 ma), Early Jurassic (Pliensbachian, 190 ma), Middle Jurassic (Bathonian, 164 ma), Late Jurassic (Kimmeridgian, 153 ma), Early Cretaceous (Aptian, 118 ma), and Late Cretaceous (Maastrichtian, 70 ma). Note the gradual reorganization of the continental masses relative to one another.

America—that's where most paleontologists live. In fact, six countries, each with a long tradition of dinosaur paleontology, account for about 75 percent of all known dinosaur fossils (Argentina, Canada, China, England, Mongolia, and the United States).[3] As the list indicates, exploration in the northern hemisphere is far ahead of that in the southern hemisphere. Approximately 75 percent of all known dinosaur species have been found on northern "Laurasian" continents.[4] As exploration of the southern hemisphere catches up, so will our knowledge of these more southerly dinosaurs.

2) Rocks of the correct age (Late Triassic through Late Cretaceous) are not preserved or exposed in some areas of the world. Sometimes the rocks simply don't exist—perhaps they were never deposited, they were eroded away, or they are buried under younger rocks. At other times, the correct-age rocks exist, but aren't exposed at the surface due to soil and vegetation cover. For this reason places like the dry, arid badlands of western North America are rich in dinosaur fossils, while places like Ecuador, covered in forests, appear fossil-free. As paleontologist Peter Dodson once remarked, "England would be one of the richest dinosaur areas in the world if it wasn't for all those trees covering them up."

A good example of the vagaries of rock preservation can be seen by examining the dinosaur distribution list in Table 1. Only a few Middle Jurassic dinosaurs have been found, particularly when compared to the many dinosaurs from the Late Cretaceous, because very few fossil-bearing Middle Jurassic rocks are exposed anywhere. Any new Middle Jurassic dinosaur is an exciting find by virtue of its rarity.

3) Some ancient environments do not fossilize. For example, "upland" environments, such as mountains, are constantly eroding away, rather than depositing sediments and fossils. Dinosaurs that lived in mountainous regions may never be found, because the likelihood they were preserved as fossils is very low.

These are just three of the main reasons for gaps in the fossil record—many others exist. No matter how hard we search the world for fossils, there will always be gaps in our knowledge of dinosaurs and their distributions. De-

spite this sobering realization, we can still learn much regarding dinosaur biology, evolution, and paleobiogeography from what we do know.

But Which Dinosaurs Are We Talking About?

The dinosaur distribution list in Table 1 tells us where and when dinosaurs have been found around the world, but not which dinosaurs. We can see that dinosaurs existed around the globe—but not all species or sub-groups of dinosaurs lived everywhere in every time period. Remember that throughout the Late Triassic, Jurassic, and Cretaceous Periods there was a continuous turnover of species as new dinosaurs evolved and others went extinct; those living in the Middle Jurassic are not the same as those in the Late Cretaceous. Knowing exactly which dinosaurs are represented at each locality and time period adds another dimension to studying dinosaur distributions. With this information, paleontologists can attempt to trace dinosaur distribution as each group and species of dinosaur evolved through time and moved through space.

Distribution patterns for specific dinosaurs and dinosaur groups can sometimes be traced through time. Some dinosaur species and groups appear to have been restricted to relatively small areas and time periods, while others appear to be quite widespread and long lived. For example, ceratopsids (the big horned dinosaurs) are known only from the western part of North America in the Late Cretaceous. But primitive iguanodontians— large bipedal herbivores that are closely related to the hadrosaurs (the duck-billed dinosaurs)—are known from every continent but Antarctica, and existed from the Late Jurassic through the end of the Cretaceous. And not every member of a dinosaur group necessarily behaved like all the others. For example, nearly all genera of hadrosaurs are restricted to a limited geographic area. But the hadrosaur genus *Saurolophus* had a relatively broader range, living in both western North America and Asia.

The distributions of longer-lived species and groups can change through time. For example, some protoceratopsians (small "precur-

sors" to the larger horned dinosaurs) occur in Early Cretaceous age rocks in Asia. However, by the latter half of the Cretaceous Period, protoceratopsians can be found in both Asia and North America. By tracking protoceratopsian distributions through time, paleontologists surmise that these small dinosaurs originated in Asia and later spread east via the Bering Strait into North America.

By tracking where each particular dinosaurs is found, paleontologists develop hypotheses of where groups may have originated, the extent of their range, and where they may have moved to expand their range. Studying the movements—or more technically the dispersal—of dinosaur species and sub-groups to new areas requires knowledge of the geography of the Earth through time. After all, dinosaurs needed to get from one place to another by foot. We need to be aware that as the dinosaurs evolved and moved around through the Mesozoic, the Earth was undergoing its own revolution in shape and configuration (Fig. 1). Knowing the history of the Earth provides the final element to any study of dinosaur paleobiogeography.

Navigating Through A Changing World

The Mesozoic world was very different from that of today. Throughout the Mesozoic, continental and oceanic connections changed, the proportion of land to sea altered, mountains were thrown up and eroded, continents shifted their positions, and the shape and size of oceans altered. Changes in the physical world have the power to drastically alter crucial environmental/climatic factors as well, including amount of rainfall, sea level, temperature range, oxygen and carbon dioxide concentrations, and oceanic circulation patterns. Many of these changes were driven, as they are today, by plate tectonic processes deep in the Earth. These processes were in full swing throughout the reign of the dinosaurs.

The configuration of the continents and oceans through the ever-changing Mesozoic world is shown in Fig. 1. Not only was each of these time periods different from today, they

were also different from one another. Dinosaurs originated in a time when all the continents were joined as the single supercontinent called Pangea (Late Triassic). Slowly Pangea broke into two continents, Laurasia in the north, and Gondwana in the south. By the close of the Mesozoic (Late Cretaceous), Laurasia and Gondwana had fragmented into the smaller continents we know today and had moved to approximately their present positions.

One can readily understand how continental rifts (separations) and collisions affected dinosaur movement. If the connection between two continents is severed (such as when Africa and South America were divided in the Early Cretaceous by the newly formed Southern Atlantic Ocean), dinosaurs would no longer be able to walk freely between these landmasses. The dinosaurs on each continent would become isolated from one another, and evolve along separate pathways. If continents which where once separated join together (such as when India collided with southern Asia shortly after the Cretaceous), disparate faunas could mix together. This situation could spell disaster for some species, and be a great boon for others.

Other changes in the Earth, like small- to large-scale climate change, can be somewhat more difficult to investigate, but equally important. As mentioned earlier, protoceratopsians and the duck-billed *Saurolophus* were able to travel between Asia and North America along the Bering Strait land bridge. The Bering Strait likely became emergent (above water) sometime in the Cretaceous when global sea level dropped. Lowered sea level allowed this shallowly submerged land bridge to become dry land. In this case, neither Asia nor North America moved in relationship to one another, yet other Earth factors caused a sea level change that united, if only temporarily, these large landmasses.

Knowledge of global climates and environments of the past can assist paleontologists in paleobiogeographic studies. Paleoclimatic and paleoenvironmental maps can be reconstructed by studying climatic indicators gleaned from the rock and fossil record (particularly from

plant fossils). Data from the Late Jurassic indicated a broad band of desert cutting across the center of both South America and Africa (Fig. 2).[5] So, while the northern and southern parts of these continents were physically connected, dinosaurs that were not adapted to desert conditions were probably restricted from migrating across such a dry, arid expanse. This example

cause the rocks indicate a desert environment, the dinosaurs that lived there must have been adapted to these dry conditions.

The shape of the land mass can also be important. Interestingly in Romania, Late Cretaceous dinosaurs (including sauropods, hadrosaurs, and ankylosaurs) all appear unusually small in size. Searching for a reason for this

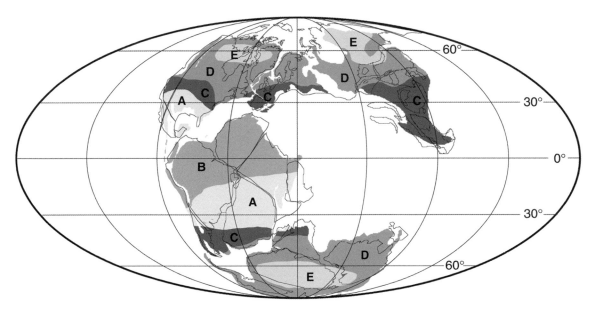

Fig. 2. Continental configurations and climatic ("biome") reconstructions for the Late Jurassic Period, approximately 150 million years ago. Note the broad band of desert slicing through the centers of what are now South America and Africa. The five major biomes are desert (A); subtropical/seasonally dry but wet in summer (B); seasonally dry/wet in winter (C); warm temperate (D); and cold temperate (E). These biome distributions are based on climatic indicators from the fossil and rock record. Modified from Rees et al. (2000).

shows that physical connections may not be enough to allow large-scale movements of dinosaurs to occur—the conditions on the Earth's surface were equally important.

Lifestyles of the Big and Scaly

Sometimes the precise place we find dinosaur fossils provides clues to their lifestyles. Commonly the clues are found in the rocks themselves, revealing secrets of the local environment and climate. The Djadoktha Formation in Mongolia, the nature of the rock itself provided clues regarding the dinosaurs' lifestyles. These Late Cretaceous rocks consist of fossilized sand dunes that entombed dinosaurs, such as *Protoceratops,* and *Oviraptor*.[6] Be-

miniaturization of Romanian dinosaurs, Dr. David Weishampel and his colleagues examined the available paleogeographic evidence and noted that in the Late Cretaceous Romania was an island. They knew that dwarfism (a reduction in size) is a common phenomenon seen in animals on small, isolated islands. These paleontologists proposed that this is exactly what happened to the Romanian dinosaurs—their small size evolved as a result of "island dwarfism."[7]

A multitude of small Early Cretaceous dinosaurs has been discovered in Australia. Paleontologists think that these dinosaurs lived between 70 and 85 degrees south latitude, placing them well within the Antarctic Circle.[8]

While the thought of polar dinosaurs seems impossible by today's climate standards, many scientists think that the world was warmer back then, and the environment these dinosaurs lived in was humid and cool, but not cold. Nevertheless, dinosaurs had to contend with long periods of dark and light, just as polar regions experience today. While examining the inside of the brain case of one polar dinosaur—a small, bipedal ornithopod called *Leaellynasaura*—Drs. Tom and Pat Rich discovered that it appears to have very large optic lobes, the part of the brain that processes visual information. They suggested that *Leaellynasaura* evolved a greatly heightened sense of vision, which was necessary during the long, dark days of the polar winter. Although speculative, this observation shows how information from many quarters—paleontological, paleoenvironmental, and geological—can be combined to form hypotheses concerning dinosaur lifestyles.

Putting All the Information Together

The who, where, and when of dinosaur distributions provide raw data for delving deeper into dinosaur evolution, dispersal, and lifestyle. Analyzing all of these patterns and effects requires much cooperation between scientists who study ancient climates, geology, and paleontology. Although dinosaur fossils themselves reveal a wealth of information, knowledge of their paleogeography and paleoenvironment puts them into a context that allows further questions to be asked and answered.

References

1. Weishampel, D.B. 1990. Dinosaur distributions. In *The dinosauria*, eds. D.B. Weishampel, P. Dodson, and H. Osmolska, 63–140. Berkeley: University of California Press.

2. Flynn, J.J., M. Parrish, B. Rakotosamimanana, W.F. Simpson, R.L. Whatley, and A.R. Wyss. 1999. A Triassic fauna from Madagascar, including early dinosaurs. *Science* 286:163–765.

3. Holmes, T., and P. Dodson. 1997. Counting more dinosaurs—how many kinds were there (1996)? In *DinoFest International Proceedings*, 125–128.

4. Forster, C.A. 1999. Gondwanan dinosaur evolution and biogeographic analysis. *Journal of African Earth Sciences* 28:169–185.

5. Rees, P.M., A.M. Ziegler, and P.J. Valdes. 2000. Jurassic phytogeography and climates: new data and model comparisons. In *Warm climates in Earth history*, eds. B.T. Huber, K.G. Macleod, and S.L. Wing, 297–318. Cambridge: Cambridge University Press.

6. Loope, D.B., L. Dingus, C.C. Swisher, III, and C. Minjin. 1998. Life and death in a Late Cretaceous dune field, Nemegt Basin, Mongolia. *Geology* 26:27–30.

7. Weishampel, D.B., D.B. Norman, and D. Grigorescu D. 1993. *Telmatosaurus transsylvanicus* from the Late Cretaceous of Romania: the most basal hadrosaurid dinosaur. *Palaeontology* 36:361–385.

8. Rich, T.H.V., and P.V. Rich. 1989. Polar dinosaurs and biotas of the Early Cretaceous of southeastern Australia. *National Geographic Research* 5:15–53.

Revealing the Whole Picture: Reconstructing an Ecosystem

David E. Fastovsky
Department of Geosciences
University of Rhode Island, Kingston

David E. Fastovsky
is a Professor of Geosciences at the University of Rhode Island. He received a B.A. in biology from Reed College, an M.A. in paleontology from the University of California, Berkeley, and a Ph.D. in sedimentology from the University of Wisconsin-Madison. His research focuses on the relationship between Mesozoic vertebrates, particularly dinosaurs, and the paleoenvironments in which they lived.

One of the great attractions of vertebrate paleontology is finding new stuff. You go to exotic places, find bones, lug them back to the museum, and then you begin to discover whole lost worlds...or do you? Sure, you found some large or small or nasty or weird creatures, with the biggest this or the smallest that, something nobody has ever seen, but does this really give you meaningful insights into lost worlds? So you found a dinosaur. It was big. It walked around on land. It ate plants. But is that all you'd ever want to know about the world that spawned this creature and in which it lived? How well do you know the animal if you know nothing about its environment?

As the latter part of the 20th century rolled along, it became clear to paleontologists that just describing new and bizarre animals was not enough; in fact, it was not necessarily even intellectually satisfying. The imagination craved an environment or a setting in which to place these creatures. Virtually limitless questions popped up: Was it a rain forest? Were there mountains? Was it hot? Was it cold? Who else lived with the dinosaurs? Who ate whom (and what)? How did they mate? Did it look like any place that we can visit now? And were there always volcanoes in the background?

Vertebrate paleontologists have long recognized that these are some of the most interesting and most difficult questions in paleontology. Thus, the science of paleoecology, that is, the study of ancient ecosystems, developed. Paleoecology has a particular relevance now. If we can understand the behavior of ancient ecosystems, perhaps as we modify our own ecosystems (through pollution, deforestation, and global warming), we can get a feel for how ecosystems will respond to these kinds of changes.

How can we reconstruct an ecosystem that is long defunct? Paleoecological conclusions tend to be circumstantial, but it is hard to find fossils of behavior.[1] In the case of vertebrates, there are the remains of the animals: bones.

> *In a time of potential global warming and disruption of well-established ecosystems, understanding paleoecosystems may provide the key toward understanding how today's ecosystems will respond . . .*
>
> *David Fastovsky*

But how the bones got there is another question. In most cases the animals didn't die where they are found. Rather, they died, and then a variety of things could have happened to their remains (see Fiorillo, page 109). Most likely the carcasses rotted or were scavenged. In very rare circumstances, however, bones are preserved as fossils. Even so, the bones are rarely preserved where the animals died. We can obtain another insight into ecosystems through isolated footprints and trackways. For once in paleoecology, these do represent behavior (see Wright and Breithaupt, page 117)! Unlike their bones, footprints are exactly where the dinosaurs put them. So they supply completely different information.

Because we ourselves are vertebrates, we commonly think about ecosystems in terms of vertebrates. But in the terrestrial realm, where we—and the dinosaurs—live(d)—most of what gives a particular ecosystem its character is the plants and invertebrates, in particular, the insects. Also, in terrestrial ecosystems, significant nutrient recycling—storage, processing, and retrieval—is accomplished in the soils by bacteria. Achieving an understanding

of ecosystems means an understanding of these facets of the ecosystem, as well.

An attempt to reconstruct paleoecology must incorporate data from several diverse fields. A source of information commonly overlooked, but key in understanding paleoecology, is (a) the sedimentary environment in which the bones are found. Of course, the bones are now in rocks, but at one point—when dinosaurs were walking around—these rocks did not exist. They were formed from sediments that were being deposited in a sedimentary environment, such as a river floodplain, or perhaps along a lakeshore, or even at a beach. These sediments covered any dinosaur skeletal remains that were present and eventually, after burial, the buried sedimentary environments (and the bones they contained) turned to rock. These preserved ancient environments, among many, many others, are recognizable to sedimentologists (geologists who study sedimentary environments) and they can be reconstructed. Also significant in terms of understanding ancient terrestrial environments, paleoenvironments, are (b) ancient soils—the substrate where organisms lived and grew. And finally, (c) climates have an effect on ancient environments.

Instead of focusing on particular environments and particular dinosaurs (which, with 160 million years of dinosaurs on Earth, would easily fill several books like this), this small chapter on dinosaur paleoecology will attempt to address aspects of large-scale changes on Earth during the years when dinosaurs were on the scene.

Dinosaurs appeared approximately 228 million years ago (mya) and the non-avian forms went extinct more or less 65 mya. Viewed from a modern perspective, this was a significant time in Earth history. As a first pass at un-

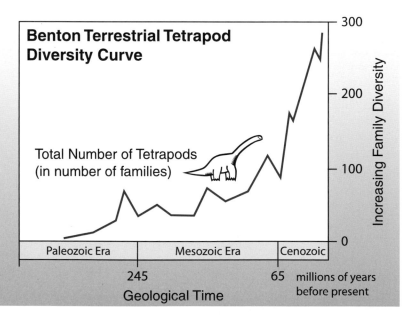

Fig. 1. The Benton curve, which represents a tally of the families of all known terrestrial tetrapods—animals with four legs—through time, shows clearly that vertebrate diversity is increasing. However, this increase may be due to a combination of factors external to organic evolution, as well as to organic evolution. From Benton (1999).[2]

derstanding ancient ecosystems, we need to gain insights into a few of the important events of those years. Then, we can overlay our understanding of dinosaur evolution during that time, and see if it is possible to recognize potential causes and effects. *Note: For the rest of this chapter, the term dinosaur will refer to the non-avian forms.*

Dinosaur Diversity and Migrating Continents

We start by looking at land-vertebrate diversity. In essence, we ask ourselves how many types of land vertebrates there were and when they lived. Among the diversity compilations, Michael Benton's is one of the best (Fig. 1).[2] Most importantly for our purposes, it shows that the diversity of vertebrates has been generally increasing over time. Diversity could appear to be increasing over time because as we look closer and closer (in time) to the present, there are more and more rocks that preserve vertebrate remains. The rocks behave like the memory: we remember better what happened more recently and less well what happened longer ago. One way to learn whether or not the increase in vertebrates is solely due to preservation is to compare the number of localities that produce vertebrates with the pattern of increasing vertebrate diversity. If indeed there are more localities as one looks at younger and younger time slices, then perhaps the increase in vertebrate diversity is at least partly a function of the number of vertebrate-bearing localities.

Suppose that the increase in diversity was not solely due to how close we look to the present; suppose that it were instead due to something else external to vertebrate evolution. One interesting factor to consider is that of continental distributions. In the time during which dinosaurs were the dominant vertebrates on land, continents became more evenly dis-

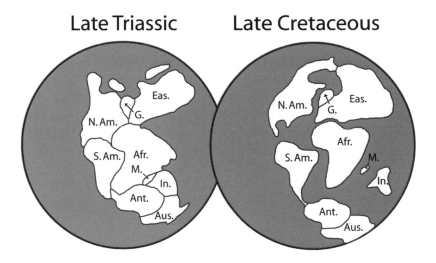

Fig. 2. Continental distributions on Earth at the beginning (Late Triassic) and end (Late Cretaceous of the Age of Dinosaurs. Whereas the Late Triassic was a time in which the continents were coalesced into a single supercontinental mass, the appearance of the continents in Late Cretaceous time was reasonably close to the way it is today. Abbreviations: Afr., Africa; Ant., Antarctica; Aus, Australia; Eas., Eurasia; G., Greenland; I., India; M., Madagascar; N. Am., North America; and S. Am., South America. Redrawn from Lillegraven et al. (1979).[7]

tributed around the globe. A look at the continental distributions during the Late Triassic shows that the supercontinent Pangea had just begun to separate (Fig. 2). Towards the end of the reign of dinosaurs, the continents had assumed positions not unlike those in which we find them today. Thus, the 160 million years of dinosaurs on Earth were characterized by increasing separation of the continents.

Interestingly enough, this separation would dramatically affect dinosaur diversity. You can imagine that if there were no barriers to gene flow, populations would remain fairly homogenous. However, if barriers appeared to isolate a small group of organisms genetically, this group and the parent population from which it was isolated could potentially evolve in different directions.

On a large-scale, this isolation is exactly what appears to have occurred during the Mesozoic. When the continents were closely linked, dinosaur diversity was at a relatively low level. Indeed, a globally homogenous vertebrate fauna existed throughout the Late Triassic–Early Jurassic interval. The dinosaurs that characterized these faunas were theropods and prosauropods, as well as some small, primitive ornithopods.

By an apparent quirk of fate (for no other explanation exists), very few Middle Jurassic localities containing terrestrial sediments have been preserved. However, considerable evidence suggests that the Middle Jurassic was a time of significant radiation of dinosaurs and other vertebrate faunas. As the continents became separated and seaways expanded between adjacent continental masses, there appeared just the type of natural barriers to gene flow already described. By the Late Jurassic, a variety of dinosaur groups appeared that had not been seen before. These groups included stegosaurs, ankylosaurs, advanced ornithopods, and sauropods, not to mention new forms in groups that were already well known, such as theropods. The evolution of such groups probably was initiated during the Middle Jurassic. Although we don't often think of evolution in non-biological terms, the physical movements of the continents clearly contributed to the distinctive pattern recorded in Benton's diversity curve (Fig. 1).

Continental movement is one example of an extrinsic factor affecting evolution. Now suppose that Benton's pattern of increasing diversity were due to something intrinsic to vertebrate evolution. If over time vertebrates found new ways to exploit their environments, then their diversity could increase through innovations: new biological inventions that would allow them to exploit resources previously unavailable to them. To understand this idea further, we need to look at plant diversity during the Mesozoic.

Plant Diversity During the Age of Dinosaurs

Plants are the primary source of energy in terrestrial ecosystems. "Primary source" means that everything else depends upon plants: the herbivorous animals that eat the plants, the carnivorous animals that eat the herbivores; the omnivores that eat both, and the scavengers that eat whatever they can. Because plants are so important, it is to be expected that the evolution of plants would have ramifications for the evolution of the vertebrates that eat them. In the case of dinosaurs, this certainly appears to have been the case.

Fig. 3 documents the rise of these various groups of plants during the age of dinosaurs. During the Late Triassic, a group of plants called gymnosperms underwent a tremendous increase in abundance. Gymnosperms had a new innovation of their own—the seed—that made dispersing them across landscapes vastly more efficient. We can think of seeds as little capsules full of nutrients to help very young plants get started. Seeds are clearly a very nutritious part of the plant and are attractive to herbivores.

The group of gymnosperms called conifers increased dramatically in the Late Triassic. Living conifers include familiar plants like pine trees. Conifers are distinctive plants, with a tendency towards great height—some conifers from the Petrified Forest in Arizona reached as high as 60 ft (20.9 m)—and very coarse, woody trunks, perhaps to support that growth. Conifers also tend to produce a lot of complex chemicals that usually taste bad or are poisonous. Considered on the face of it, this suite of adaptations is interesting: great height; chemicals that taste bad or cause sickness; and very coarse, woody, non-nutritious trunks. All of these attributes may have arisen to enable conifers to protect themselves—and their seeds—from being eaten.

A consequence of the rise of conifers was a distinct tiering—at multiple levels—of plants on Earth. On the one hand, there was a lot of low vegetation (this was not new); on the other hand, there were now very tall plants with nutrient-rich seeds. Groups of animals that were to be successful needed to be able to exploit both resources.

About 60 million years later, in the Early Cretaceous, a new type of gymnosperm appeared on Earth. These were the angiosperms—plants that bear flowers. Angiosperms showed a whole new kind of adaptation—modified leaves that attracted attention and became an extremely effective mechanism for seed dispersal. Angiosperms appear to differ from conifers in their approach to predation. Far from avoiding being eaten, they apparently encourage it, with well-developed, tasty fruits that envelop the seeds. We, of course, eat the fruits and avoid—or void—the

seeds. Thus, we (and many other omnivores) are used by the plants as a means of seed dispersal. Angiosperms, although first appearing in the Early Cretaceous, underwent a dramatic rise in diversity through the 70 million years of the Cretaceous, a rise in diversity that continued for several million years after the end of the Cretaceous.

Dinosaur Innovations

At the same time as these various plant groups were evolving, dinosaurs were evolving new herbivorous forms. Was that evolution a response to the plants, or was the plant evolution a response to the dinosaurs? This question is a chicken-and-egg debate, for which we will never have a definitive answer. But the record is clear enough that both groups were evolving strategies in response to each other.[3, 4] Fig. 3 also shows the first appearances and durations of various groups of dinosaurs.

Let's consider some of the earliest dinosaurian herbivore faunas: primitive ornithischians called ornithopods and primitive saurischians called prosauropods (see Holtz, page 31; Dodson, page 153). The ornithopods were small bipeds—perhaps three feet of body and another three feet of tail. The prosauropods, on the other hand, were large animals the largest exceeded 30 feet in length, including tail. The mouths of prosauropods were clearly capable of reaching into high vegetation—just as the early ornithopods must have restricted their foraging to comparatively low-level plants.

These two groups of dinosaurs may be viewed as partitioning the vegetation available in the Late Triassic

and Early Jurassic into plants accessible to low- and high-browsing herbivores, e.g., tiers. This kind of resource partitioning was an extraordinary innovation. Although mammals certainly do it now (think of giraffes, horses, and sheep, to provide an example of three different heights of mammalian vegetation consumption), partitioning had never been done before. It allowed dinosaurs to exploit resources that— if they existed—were heretofore unavailable to terrestrial vertebrate herbivores.

The Middle to Late Jurassic time interval, as we have seen, was a time for the establishment of conifer dominance in terrestrial ecosystems. Interestingly enough, this was also the

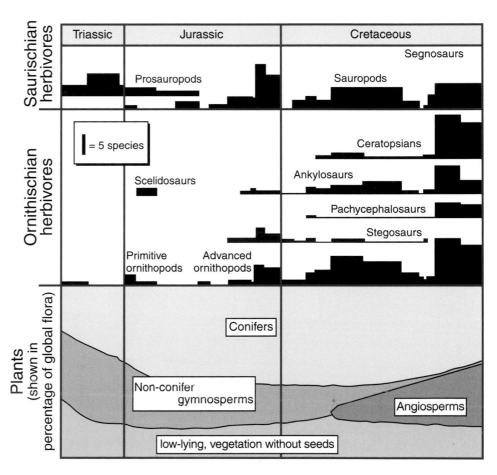

Fig. 3. Abundance of a variety of major groups of plants contrasted with the time ranges for various groups of herbivorous dinosaurs. Time and diversity are arranged as in Fig. 1. Note the significant increases in the abundance of conifers as the Mesozoic progresses. Note, too, the appearance and rapid radiation of the angiosperms in the Cretaceous. Compare these observations with the variety of new herbivorous dinosaur types that appear in or just before the Cretaceous, perhaps reflecting some relationship to the diversification of plants at that time. Redrawn from Fastovsky (2000);[8] data from Weishampel and Norman (1989) and Tiffney (1997).[9, 10]

time that sauropods—herbivorous dinosaurs of extraordinarily large size with long necks (see Dodson, page 153)—probably first made their appearance. Again, we say "probably," because we know so little about the terrestrial Middle Jurassic. However, the rise of sauropods appears to have been nearly lockstep with the rise in abundance of conifers, and it is probably not too outrageous to suggest that these two groups of organisms evolved together to large sizes and heights as a result of their mutual interactions, e.g., sauropods eating conifers.

By the Late Jurassic, sauropods were a diverse and abundant group, and conifers were clearly the most abundant large plant type on Earth. A scientist, R.T. Bakker, has suggested that dinosaurs "invented" flowering plants in the Late Jurassic.[5] He reasoned that the mighty sauropods overgrazed the conifers and this overgrazing allowed flowering plants to evolve on denuded landscapes in the Early Cretaceous. While no evidence supports this theory, the relationship between dinosaurs and plants might hold the answer to a question that has plagued students and dinosaur-enthusiasts since dinosaurs were discovered: why did they grow so large? If in fact sauropods grew to large sizes at least in part to take advantage of tall conifers, it may be that dinosaurs owe their large size to plants. For, as sauropods grew to extreme sizes to feed from the tallest conifers, the carnivorous dinosaurs that must have fed upon them may also have increased in size to manage their large prey. Considered this way, plants and plant-eaters may have been in an ever-increasing size race: the plants grew taller to avoid the sauropods, which themselves grew larger to reach high vegetation as well as protect themselves from large carnivores. The large carnivores grew larger to more effectively handle the large herbivores. And so it went, until these organisms reached near maximum sizes for life on land.

By the Cretaceous Period, certain groups of herbivorous dinosaurs developed remarkable adaptations for handling plants. A bit of background is necessary here: Most mammals chew their food, breaking it down in the mouth before sending it rearward for further digestion. However, chewing is really quite unusual among vertebrates. Most vertebrates only use their mouths to grab food, and—in a few groups—chop it up into chunks which are in turn sent rearward. Where does the heavy-duty grinding and enzymatic breakdown take place? Not in the mouth! The real work is done in the gizzard, a muscular part of the stomach that may contain stones (swallowed by the animal). The stones are a kind of natural mortar and pestle that the animal uses to grind down food. For this reason we sometimes see birds "eating" stones, and the remains of dinosaurs are occasionally found with polished stones within the rib cages. These stones, called gastroliths, are the gizzard stones with which the animals ground their food.

A look around the living animal kingdom would suggest that chewing is one of the great inventions of mammals. It turns out that dinosaurs also chewed. However, they developed this adaptation independently of mammals. In the Cretaceous, just as the angiosperm revolution was getting underway, hadrosaurs and ceratopsians began to develop fantastic chewing innovations. The skulls of hadrosaurs and ceratopsians had a number of key chewing features very similar to those of herbivorous mammals, including

- Cropping beaks at the front of the jaw for biting off vegetation;
- Blocks (or shearing masses) of tightly packed molar-like teeth for grinding or cutting vegetation;
- Cheeks for keeping the food in the mouth;
- Robust, well-developed coronoid processes (expansions at the back of the lower jaws) for the attachment of strong muscles for jaw closure.

Both groups—ceratopsians and hadrosaurs on the one hand, and modern herbivorous mammals on the other—evolved very efficient ways of processing food in the mouth. The paths that led them to similar end points are quite disparate. Dinosaurs developed many of these "mammalian" chewing characteristics some 30 million years before mammals got around to "inventing" them!

Because the evolutionary paths that led to these similarities in chewing are so different, it

is to be expected that there would be differences between the mammalian chewing apparatus and those of ceratopsians and hadrosaurs. Indeed, the more closely we look, the more striking are the differences we see. For example mammalian teeth are shed once (baby teeth), and then adult teeth take their place for the rest of the animal's life. In the case of a horse, which can live to 30 or more years, this can be a problem. Horses grind their molars against silica-rich grasses, and after many years of this abrasion, the enamel tops of the teeth are utterly worn down. To compensate, horses develop teeth with very tall enamel tops, i.e., high-crowned teeth, but in long-lived individuals, this may not be enough. Hadrosaurs and ceratopsians solved this problem a very different way. They simply replaced teeth constantly as they wore out (in the manner of other reptiles). Mammals develop very close fits—occlusion—between upper and lower teeth, so that the grinding is most efficient; dinosaurs never really developed tight occlusion.

Significant differences exist between the ways that ceratopsians and hadrosaurs developed chewing, as well. By looking at striations produced as the upper teeth wore against the lower, we can tell that ceratopsians had an almost vertical slicing plane (the striations literally run down the side of the teeth), whereas that of hadrosaurs was closer to horizontal (the striations run at an angle across the tops of the teeth.[6] This means that ceratopsians were effectively far less efficient chewers than were hadrosaurs. Does this say anything about the respective behavior of these groups? It is tempting to speculate that hadrosaurs, with their magnificently developed chewing apparatus, easily ground through the tall, woody conifers that were so abundant throughout the Cretaceous. Ceratopsians, on the other hand, are a Late Cretaceous, mainly North American group (all the large ones were North American). Their rise occurred at more or less the same time as the angiosperms. Can it be that ceratopsians, with an apparently less efficient chewing mechanism than hadrosaurs, were restricted to eating the blossoming—and more chewable—angiosperms, while the hadrosaurs restricted themselves to the conifers?

Part of the puzzle is clear enough: stomach contents of hadrosaurs have been found and these contained coniferous materials. Likewise, fossilized feces (coprolites) attributed to hadrosaurs have been found, and these contained conifer material as well. But what of the ceratopsians? This remains unknown. But even in the Late Cretaceous, we find a hint that these two abundant groups of herbivorous dinosaurs may have been partitioning the food resource as well—not only by height—which continued to be an effective way to distribute the resource—but also by vegetation type. These speculations are hypotheses that we hope to test in the future, perhaps by finding coprolites whose owners can be identified, or perhaps by the observation of a clear relationship between particular groups of dinosaurs and particular plants.

Conclusions

It is tempting to think of dinosaur studies as collecting bones and reconstructing them into impressive museum displays. But we can do much, much more, and in a way that may be as interesting—or more interesting—than the bones themselves. Using the techniques of sedimentary geology and taphonomy, it is possible to reconstruct the environment in which any given dinosaur lived or died (see Fiorillo, page 109). We know that dinosaurs not only lived in lush tropical environments (such as those in which they are often depicted), but also more commonly lived in places like arid deserts, woody forests, along riverbanks, in the high Arctic…indeed, just about anywhere on Earth. As we have done in this chapter, we can also look at the big picture of dinosaur and plant evolution, and see dinosaurs apparently responding by diversification to the evolution of a variety of plant types on Earth.

The story has hardly begun to be told. New correlations and a refined understanding of plant and dinosaur distributions will be keys not only to testing the ideas outlined here, but to proposing and refining our understanding of the great dance between plants and the herbivorous dinosaurs that ate them. Paleoecology, therefore, gives insight into the way that eco-

systems work over periods of time far longer than we could ever observe during a human lifetime. For this reason, paleoecology is critical in studying large-scale ecosystem changes on Earth. In a time of potential global warming and disruption of well-established ecosystems, understanding paleoecosystems may provide the key toward understanding how today's ecosystems will respond over time durations longer than the written record of human beings.

References

1. Fastovsky, D.E. (in press). Dinosaur paleoecology. In *The dinosauria*, 2nd edition, eds. D.B Weishampel, P. Dodson, and H. Osmólska. Berkeley: University of California Press.

2. Benton, M.J. 1999. The history of life: large databases in palaeontology. In *Numerical palaeobiology: computer-based modelling and analysis of fossils and their distributions*, ed. D.A.T Harper, 249–283. Chichester and New York: John Wiley & Sons.

3. Wing, S.L., and B.H. Tiffney. 1987. The reciprocal interaction of angiosperm evolution and tetrapod herbivory. *Annual Review of Paleobotany and Palynology* 50:179–210.

4. Barrett, P.M. and K.J. Willis. 2001. Did dinosaurs invent flowers? Dinosaur-angiosperm coevolution revisited. *Biological Reviews* 76:411–447.

5. Bakker, R.T. 1986. *The dinosaur heresies*. New York: William Morrow and Company.

6. Dodson, P. 1996. *The horned dinosaurs*. Princeton: Princeton University Press.

7. Lillegraven, J.A., M.J. Kraus, and T.M. Bown. 1979. Paleogeography of the world of the Mesozoic. In *Mesozoic mammals: the first two-thirds of mammalian history*, eds. J.A. Lillegraven, Z. Kielan-Jaworowska, and W.A.Clemens, 277–308.

8. Fastovsky, D.E. 2000. Dinosaur architectural adaptations for a gymnosperm-dominated world. In *Phanerozoic terrestrial ecosystems*, eds. R.A.Gastaldo and W.A. DiMichele. *Papers of the Paleontological Society* 6:183–207.

9. Weishampel, D.B., and D.B. Norman. 1989. Vertebrate herbivory in the Mesozoic: jaws, plants, and evolutionary metrics. In *Paleobiology of the dinosaurs*, ed. J.O. Farlow. Geological Society of America Special Paper 238:87–100.

10. Tiffney, B.H. 1997. Land plants as food and habitat in the Age of Dinosaurs. In *The complete dinosaur*, eds. J.O. Farlow. and M.K. Brett-Surman, 352–370. Bloomington: University of Indiana Press.

Acknowledgments

I thank those who reviewed this chapter. Their comments were very constructive and helped to strengthen this paper.

The Supporting Cast: Dinosaur Contemporaries

Nicholas Fraser

Virginia Museum of Natural History
Martinsville

8

Nicholas Fraser is Curator of Vertebrate Paleontology at the Virginia Museum of Natural History and is adjunct professor of Geology at Virginia Polytechnic Institute & State University. He received his B.S. in zoology and Ph.D. in geology from the University of Aberdeen, Scotland. His research interests include terrestrial faunal change at the Triassic-Jurassic boundary and the paleobiology and phylogenetic relationships of the Sphenodontia.

Dinosaurs were an incredibly successful group of animals that unquestionably deserve the enormous attention they receive. At the same time, they also detract from other equally significant aspects of Mesozoic life. From my perspective, two major pitfalls plague popular dinosaur literature. First, is the implication that if a vertebrate is large and extinct, then it must be a dinosaur. If that hurdle is successfully cleared, the second pitfall will almost certainly rear its head! That is the portrayal of dinosaurs as the only significant terrestrial life forms living during the Mesozoic. Let's look at these two pitfalls in turn.

A Case of Mistaken Identity

In arenas of everyday life that have a direct and profound effect on children, many vertebrates that were contemporaries of dinosaurs (particularly the large sea-going and flying reptiles) are too frequently referred to as dinosaurs. No breakfast time would be complete without a gaudy plastic pterosaur popping out of a cereal box incorrectly proclaiming its identity as a dinosaur. Sadly, this error is also seen in children's books where *Dimetrodon*, *Pteranodon*, plesiosaurs, and even woolly mammoths are commonly implied to be dinosaurs. It is little wonder we grow up with a number of misconceptions about dinosaurs. As educators, we are often guilty of oversimplifying to the point of providing children with inaccurate information. Then a few years down the road, some poor soul is faced with the dilemma of telling the students that they were previously misled. I frequently hear the argument that we shouldn't tell school children that now most paleontologists regard the birds at their garden feeders simply as a specialized group of theropod dinosaurs. "It will just confuse them" is the repeated phrase in these situations. It seems to me that the reverse is really the case. Telling them one thing at one age, and then another at a later date is more likely to confuse. Let's get the most up-to-date ideas across to young students at the beginning, and if in reality the parents (and other educators)

> *. . . dinosaurs were just one of a number of very important and specialized vertebrate groups living in the Mesozoic.*
>
> *Nick Fraser*

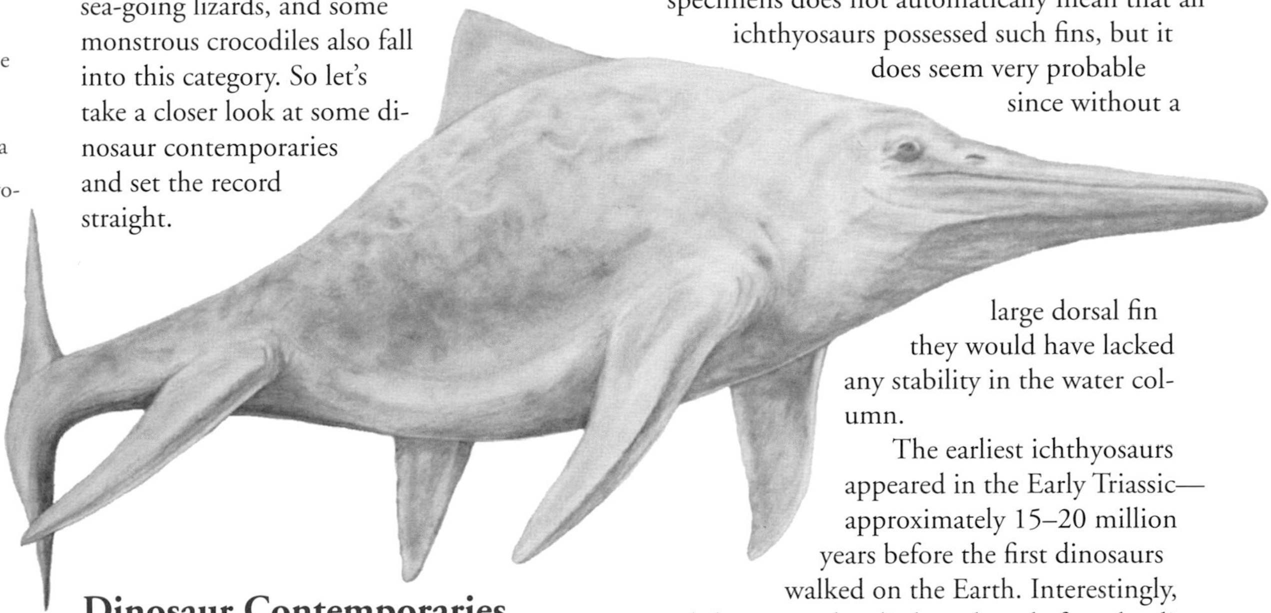

Fig. 1. Restoration of the ichthyosaur, *Shonisaurus*, illustrating the dolphin-like body plan of a typical ichthyosaur.

are the ones confused by the ever-changing face of science, we need to educate them at the same time!

In the line-up of dinosaur suspects, the principal red herrings are the dolphin-like ichthyosaurs, the ocean-going plesiosaurs, and the flying pterosaurs. On occasion, some of the large, extinct members of living groups are also misidentified. The mosasaurs, a group of giant sea-going lizards, and some monstrous crocodiles also fall into this category. So let's take a closer look at some dinosaur contemporaries and set the record straight.

that of a modern-day tuna (Fig. 1). This reconstruction is based on the fact that some specimens preserve images of the body outline as a carbon film, showing prominent dorsal and tail fins. In the past, similar body outlines were painted onto the sediment around the fossil bones, thereby giving the mistaken impression that preservation of dorsal fins was common. Of course finding dorsal specimens in some specimens does not automatically mean that all ichthyosaurs possessed such fins, but it does seem very probable since without a large dorsal fin they would have lacked any stability in the water column.

The earliest ichthyosaurs appeared in the Early Triassic—approximately 15–20 million years before the first dinosaurs walked on the Earth. Interestingly, ichthyosaurs also died out long before the dinosaurs disappeared, the last ones living approximately 90 million years ago. Perhaps they failed to compete successfully with the advanced sharks that began to radiate at that time.

Dinosaur Contemporaries

Ichthyosaurs—or "fish reptiles" (early Triassic–mid-Cretaceous)—superficially resembled toothed whales (Fig. 1). Their relationship to dinosaurs is remote indeed, lacking all the features of terrestrial adaptation and the upright posture. Indeed, some of them may well have occupied similar niches to modern cetaceans (whales). All ichthyosaurs had essentially the same body plan and it is widely accepted that they were completely unable to come out on dry land. In fact, substantial evidence exists to indicate that they even gave birth to live young at sea. A number of specimens have been found with embryos preserved still within the mother's body, and there is at least one individual that apparently died during childbirth—the baby may have stuck in the birth canal as it was born tail first.

In ichthyosaurs, both front and hind limbs are modified to form paddle-shaped organs. In addition, restorations of the animal show a large dorsal fin and a large lunate tail fin like

Some of the largest ichthyosaurs were among the earlier members of the group. *Shonisaurus* from the Late Triassic attained a length of 15 m (49 ft). Ichthyosaurs seem to have reached their peak diversity in the Jurassic, and the numerous fabulously preserved specimens from Germany and England give us a glimpse of fast moving predators, similar to the tunas and sharks of today. The fish and cephalopods of the Jurassic seas would have provided abundant prey.

Plesiosaurs (Early Jurassic–Late Cretaceous) were the Loch Ness monsters of the Mesozoic world (Fig. 2D). They can be divided into two basic groups: the long necked plesiosauroids and the shorter-necked pliosauroids with a

much more conspicuous skull. In both groups the limbs were modified to form paddles or flippers. Over the years there has been some debate concerning the way these modified limbs were used. This debate serves to show just how science works by constantly testing hypotheses and refining our ideas. It was suggested initially that plesiosaurs used their limbs as oars and that they rowed through the water (Fig. 2A). This idea was certainly very attractive given the shape of the flippers. But, although the angle could be adjusted, the flippers could not be removed from the water. Therefore, the force of the backstroke would create a counter thrust that would cancel out at least part of the forward motion.

An alternative method of movement would be the kind of underwater flight that is employed today by penguins and sea turtles. In this case, the up and down movement of the paddles would provide lift, with the flippers describing a figure eight (Fig 2B). The cross-section of the flipper, with its airfoil shape, supports this hypothesis.[1] However, an examination of the range of movements possible at the shoulder girdle indicated that plesiosaurs were incapable of significant up and down movement of the flipper. So another explanation was sought.

The answer to the problem seems to lie in a version that incorporates elements of both the flying and rowing models. Godfrey suggested that the tip of the flipper described a crescent-shaped path (Fig. 2C).[2] In this scenario, the primary propulsive force would come from the retraction of the flipper—similar to action in the rowing model. But in the recovery stroke, there would have been a component of upward lift that would have minimized the counterthrust as the limb was brought back into its original position. In fact,

this movement is precisely what is seen in living sea lions. In time, objections may be raised to this model, and it may need to be refined or even rejected. In the meantime, it is a viable explanation that is consistent with all that we currently know.

An interesting corollary of the locomotor patterns concerns the reproductive habits of plesiosaurs. While no plesiosaur specimens have been found with associated embryos, neither have any plesiosaur eggs been described. We have no way of telling what their reproductive strategy was. Nevertheless, detailed examinations of the skeletons of plesiosaurs indicate that they had very rigid trunk regions with broad plate-like limb girdles, and that little movement was possible between the individual vertebrae. If they did indeed lay eggs, it would

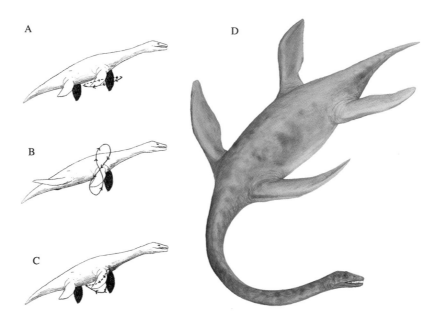

Fig. 2. Schematic diagrams to illustrate changing hypotheses regarding plesiosaur locomotion. **A.** underwater rowing, **B.** underwater "flying" like a penguin and **C.** the currently accepted intermediate method employing a flipper movements similar to a modern-day sealion. **D.** Restoration of the long-necked plesiosaur, *Elasmosaurus*.

have been a very strenuous trip up the beach, as they would have been incapable of "humping" up the beach in the manner of sea lions. Instead, they would have been forced to drag their enormous bodies across the beach in the way that sea turtles do today. A 40-foot long plesiosaur would certainly have ploughed an

impressively deep furrow in the sand—perhaps we shall find a preserved trail of one some day!

Although ichthyosaurs and plesiosaurs were large aquatic reptiles living in the Mesozoic, chances are they did not interact with dinosaurs. As far as we know, no dinosaur habitually entered the aquatic realm. When we do find dinosaurs in marine deposits, they typically occur as isolated and rather fragmentary components of assemblages rich in typical marine fossils. We can infer in such cases that the dinosaur remains were washed out to sea. In one or two instances fairly complete dinosaurs have been recovered from fully marine sediments. *Niobrarasaurus* from the Niobrara Chalk and *Scelidosaurus* from England are two examples. However, they show no definitive adaptations to an aquatic existence, and most authors regard them as fully terrestrial animals.

Pterosaurs are a group of ancient vertebrates characterized by an enormously elongate fourth finger that supported a membranous wing. In contrast with the ichthyosaurs and plesiosaurs, pterosaurs are considered to be closely related to dinosaurs, as they share a more recent common ancestor. Pterosaurs are sometimes broadly separated into two groups: the rhamphorhynchoids and the pterodactyloids. The former is something of a ragbag assemblage that is loosely characterized by an elongate tail, a relatively short neck, and a small head. The pterodactyloids appeared in the Late Jurassic and form a more natural grouping of animals. They are characterized by a much-reduced tail and elongate wrist bones. Typically, they also have much larger heads relative to the length of their bodies. During the Cretaceous some giant pterosaurs evolved. The truly monstrous *Quetzalcoatlus* had a wingspan approaching 40 feet—equivalent in size to a small airplane!

The traditional view of pterosaurs was one of wonderfully efficient flyers and gliders, which were clumsy and comical when waddling around on the ground. The reason for this image was that the wing membrane was presumed to have attached to the hind limb in a similar manner to that of a bat. Imagine yourself in a sack race—you can only push one foot so far in front of the other before you yank your support away! In a sack it is relatively easy to walk slowly, although in a somewhat ungainly fashion, but attempt to run and you quickly fall flat on your face. Some specimens of pterosaur that preserve remnants of the wing membrane suggest that it was attached near to the ankle. However, some workers have disputed this idea, and the examination of other specimens indicated that perhaps the wing was attached nearer to the trunk.[3] If this were the case, there is at least a possibility that pterosaurs were capable of a much brisker, upright walk (Fig. 3A). Indeed, evidence from three-dimensionally preserved specimens of certain pterosaur hipbones and hindlimbs does suggest that the limb might have been held in an upright position like that of the dinosaurs. Some of the most recent studies suggest that pterosaurs actually adopted a stance and gait that was somewhat intermediate between the two outlined above. Based on trackways known as *Pteraichnus*, widely thought to have been made by pterosaurs,[4] Bennett argued that the makers were neither sprawling quadrupedal forms nor bird-like bipedal animals.[5] Instead, they appear to have been walking on all fours, but with an erect posture (Fig. 3B).

These ideas are clearly opposing hypotheses, but perhaps they don't have to be mutually exclusive. While the modification of the forelimb into a wing poses some restrictions, that doesn't rule out the possibility that different groups of pterosaurs evolved different locomotor patterns. Even among a closely related group of animals such as the birds we see a great disparity of locomotor types. The waddling of penguins is a far cry from the high-speed run of an ostrich. It is at least conceivable that somewhat different locomotor patterns and postures existed among the pterosaurs. So, for the time being the question of pterosaur locomotion remains unresolved.

Although some of the smaller pterosaurs were most probably insectivorous, many pterosaurs were undoubtedly fish eaters. The Jurassic *Rhamphorhynchus* probably trawled its lower jaw through the water and pierced fish with its long, widely spaced teeth. It is likely that the

toothless *Pteranodon* also trawled the sea, but swallowed its prey whole. Probably the diversity of head and "beak" crests that evolved among pterosaurs served principally for display and recognition functions.[6] One Cretaceous form, *Pterodaustro*, bore multitudes of incredibly long and thin teeth on the lower jaw, and together they formed a sieve something akin to the baleen of mysticete whales. By taking in mouthfuls of water, *Pterodaustro* could force

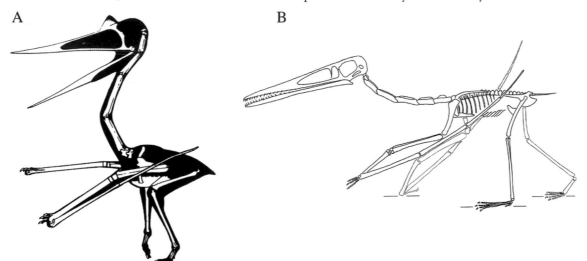

the water through the mesh of teeth and trap small marine organisms that it could then lick off and swallow.

The pterosaurs were a very diverse and widespread group. They died out alongside the last non-avian dinosaurs at the end of the Cretaceous.

The Great Supporting Cast

The giant aquatic and aerial reptiles of the Mesozoic tend to receive their quota of coverage, even if they are misidentified. But an immense array of animals are mere shadowy figures in the background. We need to recognize the critical roles that these animals played in the dinosaurs' world.

Although it is referred to as the Age of Dinosaurs, one question we might ask is how much did dinosaurs dominate the Mesozoic landscape? In a world where the biggest animals are not common everyday sightings, we should also ask whether the large dinosaurs of the Jurassic were really as obvious as many documentaries and films would have us imag-

ine. In places like Yellowstone National Park, it takes some effort to find a herd of elk or bison, and carnivores such as bears are certainly not to be found around every corner. Even on the vast open plains of the Serengeti, many large animals remain hidden from view for long periods. It's surprising what a little camouflage and secretive habits can do.

When we think of the world, either past or present, inevitably it is in a very distorted manner. Unless something has an immediate impact on us we are inclined to ignore it. It has often been remarked that it is the little things that keep the Earth turning round. For example, the plains of Africa would long since have been buried under vast piles of elephant manure if it were not for the diligent work of dung beetles. Just imagine the plight of Jurassic ecosystems without similar agents to clean up after the passage of a herd of *Diplodocus*!

The Little Guys

The scarcity of discussions concerning the terrestrial invertebrate record is a particularly glaring gap in our portrayal of the dinosaurs' world. It is unquestionably the invertebrates, and in particular the insects, that dominate terrestrial animal communities. At first it might seem natural that insects would be poorly represented as fossils. After all, a judicious swat with the hand, and a delicate-winged mosquito, complete with its piercing mouthparts, becomes nothing more than a nasty, if somewhat satisfying, smear. Therefore, how could

Fig. 3.
A. Restoration of the pterosaur *Quetzalcoatlus* in a posture where the wing membrane is largely free of the back leg (after Peters, 1989). **B.** Reconstruction of *Pterodactylus* in an erect quadrupedal posture (after Bennett, 1997).

we expect insects to fare well against the rigors of fossilization? But, despite their rather delicate nature, insects generally do have a very extensive fossil record. A Triassic locality in Virginia is producing numerous, incredibly detailed and complete insects that even exhibit the small hair-like projections on the antennae and body surfaces.[7] Together with a handful of other Triassic sites, we know that many modern day families and orders of insects, including true flies (Diptera), caddis-flies (Trichoptera), thrips (Thysanoptera), and water bugs (Hemiptera), were contemporaries of the earliest dinosaurs. What is particularly striking is the very close resemblance of these Triassic insects to the living members of their respective orders and families. For example, a fossil thrips from Virginia has the same long, narrow body and short wings with the very distinctive fringe of long hairs around their margins, as its modern day cousins. Many a blossom in a rose garden or fruit orchard has been devastated by a thrips infestation. But, in the absence of Triassic flowers, the most ancient thrips presumably plied their trade sucking the juices from alternative sources such as fungi.

Many sites in Asia show that insect groups were widespread in Jurassic times as well. And then, of course we have insects in Cretaceous amber. While such finds cannot reproduce the scene from *Jurassic Park*, they do offer evidence of the little things that made the Mesozoic world turn round. And, these little guys were surprisingly like those of our own world.

Smaller Vertebrate Contemporaries

The dinosaurs have become so much larger than life, that one incredibly important point is almost always overlooked: the beginning of the Mesozoic marked the beginning of the modern terrestrial ecosystem. It is not surprising that when we come to look at the smaller vertebrate contemporaries of the dinosaurs, most forms can be slotted comfortably into extant higher order taxa. We have the earliest mammals, turtles, crocodiles, and lissamphibians (frogs and salamanders) from the Triassic. Admittedly the Mesozoic mammals and their associates were frequently very different from their living descendants, yet they were an im-

portant part of the ecosystem, just as they are today.

Even the dinosaurs, viewed by marketing firms as the archetypal icon of the old-fashioned and out-dated, are in keeping with the modern tetrapod body plan. It is now widely accepted that birds are merely a specialized group of theropod dinosaurs. In that sense, dinosaurs can also be regarded as part of the radiation of modern terrestrial tetrapods.

Crocodiles

The earliest crocodiles were very different from the sluggish, semi-aquatic descendants that we see today. Appearing in the Late Triassic (220 million years ago), the first crocodiles were smaller, sleek, entirely land-living predators. The dog-sized *Terrestrisuchus* (Fig. 4A) is a good example of these early forms. With long, slender legs, it has even been suggested that *Terrestrisuchus* was a bipedal runner.

One curious thing about modern crocodiles is that, unlike any other living reptile, they have a fully separated four-chambered heart like that that of mammals and birds. But like all living reptiles, crocodiles are "cold-blooded." It is interesting to speculate that the four-chambered heart of today's crocodile may be a relic of its active ancestors. It is even conceivable that the earliest crocodiles were warm-blooded (endothermic). However, it must be emphasized that currently there is no way to test this idea. Because it cannot be tested and potentially falsified, this idea should be viewed as sheer speculation and currently outside the parameters of rigorous scientific methodology.

The familiar crocodile shape began to evolve fairly early during the Mesozoic. By the Cretaceous, giant forms such as *Sarcosuchus* and *Deinosuchus* had appeared. These fearsome looking creatures undoubtedly packed a powerful crunch in the jaws, and it has even been suggested that they could have preyed on large dinosaurs.[8]

Microvertebrates

In 1979 Lillegraven, Kielan-Jaworowska and Clemens edited a seminal volume dealing ex-

clusively with mammalian evolution.[9] But if you searched through the book expecting to find familiar names of ungulates (hoofed animals), rodents, whales, etc., you would be greatly disappointed. Many of the Mesozoic taxa are based on fragments of jawbones and isolated teeth, and they have unfamiliar names such as haramyids, docodonts and multituberculates. However, their obscurity in the popular literature does not diminish their importance in the Mesozoic world. It is important to know that a fossil does not have to be large and complete to be of great significance. For instance, the isolated teeth of many of the earliest mammals are so important that paleontologists are sometimes prepared to sift through tons of sediment to find a handful.[10]

Of course, compared to looking for the monstrous bones of sauropods, searching for the millimeter-sized teeth of early mammals is a bit like looking for a needle in a haystack. Their rarity only serves to enhance their value. Fortunately, the painstaking search for the remains of small contemporaries of dinosaurs has been eased through the action of millions of unwitting amateur paleontologists—namely *Pogonomyrmex* sp., the western red harvester ants! In many dinosaur-producing regions, a careful search of anthills will yield a totally new perspective on the great dinosaur beds of the west. Here the ants industriously scurry about, collecting rock grains for their nest. Their collection inevitably includes the small flakes of fossil dinosaur bone weathering away at the surface. The value of ants in locating fossil vertebrate sites has been known since the end of the 19th century when J.B. Hatcher described finding fossils with the assistance of ants.[11]

Dinosaur bone fragments are not the only vertebrate fossils encountered on these anthills. Isolated teeth and jawbones from small mammals, crocodiles, lizards, salamanders and fish can also be found. Unlike some of the dinosaur bone slivers, these fossils are often diagnostic— that is to say, they have characteristic features that can be used to identify the species from which they originated.

One group of animals that is occasionally found among the anthills is the Sphenodontia. This group of obscure reptiles is closely related to lizards, and externally they are very lizard-like. However, they have a more rigid skull structure than lizards and their robust teeth are typically fused to the edge of the jaw. It seems likely that the Sphenodontia could feed on prey with tough exoskeletons, such as grasshoppers and cockroaches. Rather than swallowing their prey whole, their robust teeth masticated them into smaller pieces before swallowing. Sphenodontians first appeared in

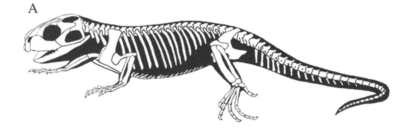

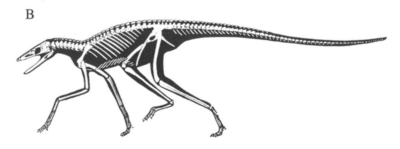

Fig. 4. Restoration of the skeletons of **A.** a typical sphenodontian (Clevosaurus) and **B.** a sphenosuchian crocodile (*Terrestrisuchus*) (a, after Fraser 1988; b, after Sereno and Wild 1992).

the fossil record at about the same time as the dinosaurs, and they flourished during the Jurassic. However, beginning in the Cretaceous, their diversity levels dwindled dramatically while at the same time, the lizards flourished. There are presently no records of sphenodontians from the Late Cretaceous through the Tertiary Period, yet there is one living member, the Tuatara (*Sphenodon*). It lives on a few isolated islands off New Zealand. The Sphenodontia is just one more example of a living group that has its origins back at the beginning of the Age of Dinosaurs.

Summary

The great dinosaur beds of the world have fascinated people for a very long time, and will undoubtedly do so for many years to come. As new finds are made and as older discoveries are examined from new perspectives, it is apparent that our understanding of the Mesozoic world is one that will continue to change. For me, two principal points stand out about the Mesozoic era, and I suspect that they will always hold true. First, dinosaurs were just one of a number of very important and specialized vertebrate groups living in the Mesozoic. Second, while the early Mesozoic is often cited as the "Dawn of the Dinosaurs," an equally appropriate slogan would be the "Dawn of the Modern World."

References

1. Robinson, J.A. 1975. The locomotion of plesiosaurs. *Neues Jahrbuch für Geologie und Paläontologie* 149:286–332.

2. Godfrey, S. 1984. Plesiosaur subaqueous locomotion: a reappraisal. *Neues Jahrbuchfür Geologie und Paläontologie* 11:661–672.

3. Padian, K. 1983. A functional analysis of flying and walking in pterosaurs. *Paleobiology* 9:218-239.

4. Lockley, M.G., J.L. Wright, W. Langston, and E.S. West. 2001. New pterosaur track specimens and track sites in the Late Jurassic of Oklahoma and Colorado, their paleobiological significance and regional ichnological context. *Modern Geology* 24:179–203.

5. Bennett, S.C. 1997. Terrestrial locomotion of pterosaurs: a reconstruction based on *Pteraichnus* trackways. *Journal of Vertebrate Paleontology* 17:104–113.

6. Bennett, S.C. 2002. Soft tissue preservation of the cranial crest of the pterosaur *Germanodactylus* from Solnhofen. *Journal of Vertebrate Paleontology* 22:43–48.

7. Fraser, N.C., D.A. Grimaldi, P.E. Olsen, and B. Axsmith. 1996. A Triassic lagerstatte from eastern North America. *Nature* 380:615–619.

8. Sereno, P.C. 2001. Super croc. *National Geographic Magazine* 200 (6):84–89.

9. Lillegraven, J.A., Z. Kielan-Jaworowska, and W.A. Clemens. 1979. *Mesozoic mammals: the first two-thirds of mammalian history.* Berkeley: University of California Press.

10. Moore, C. 1881. On abnormal geological deposits in the Bristol district. *Quarterly Journal of the Geological Society of London* 37:67–82.

11. Hatcher, J.B. 1896. Some localities for Laramie mammals and horned dinosaurs *American Naturalist* 30:112–120.

Dinosaurs as Living Animals

We cannot go to a museum and watch a fossil going about the activities of daily life. What, then, makes us think that *Maiasaura* was a "good-mother reptile," taking care of her young after they hatched? Or that *Troödon* protected its nest? **Horner**'s article describes the exciting discoveries made over the years at Egg Mountain in Montana, and explains the lines of evidence that have influenced our understanding of the behavior of several species of dinosaurs.

9

"Hot-blooded dinosaurs" make good headlines in the popular press. But, were dinosaurs— some or all of them—really endothermic? **De Ricqlès** explains how paleontologists learn about the physiology of extinct organisms, and describes the diversity of ways living vertebrates are able control their body temperatures. He presents some of the evidence for temperature regulation in dinosaurs, showing readers why we are really not asking the right question when we ask "were dinosaurs warm-blooded or cold-blooded?"

10

In 1870, not many years after the discovery of *Archaeopteryx*, Thomas Huxley announced that birds were the descendents of dinosaurs. Not everyone agreed. **Currie**'s article explains the basis for the debate about bird origins. He presents readers with possible scenarios for the evolution of feathers and flight, using recent excavations of feathered dinosaurs to illustrate how new discoveries can affect existing hypotheses.

11

12

The article by **Archibald** discusses the concept of extinction before presenting synopses of the three most popular hypotheses concerning possible causes for the mass extinction event at the end of the Mesozoic. He uses data from studies of dinosaurs to test each of these hypotheses in turn, giving readers insight into the complexity of the issues surrounding extinction studies.

Evidence of Dinosaur Social Behavior

Jack Horner

Museum of the Rockies
Montana State University, Bozeman

9

John "Jack" Horner is Regent's Professor and Curator of Paleontology at the Museum of the Rockies, Montana State University, and Senior Scholar of Vertebrate Paleobiology at the National Museum of Natural History, Smithsonian Institution. He was educated at the University of Montana where he received an Honorary Doctorate of Science. His research focus concerns the paleobiology of dinosaurs. In 1986 he was awarded a MacArthur Fellowship.

Dinosaurs are dead, extinct, and their previous existence is evidenced by little more than fossilized skeletons, skin impressions, eggs, coprolites, or traces like footprints and nests. As is true with any other extinct group of animals, it is not possible to observe their behaviors. Dinosaur behaviors have to be hypothesized from bits and pieces of morphological and geological evidence, and inferred from behaviors observed in related living animals. For dinosaurs, this means studying the behavior of their closest living relatives, the crocodilians (including alligators) and birds. If there is evidence for a particular behavior, and it's a behavior seen in crocodilians and birds, then we can make a pretty good case for it having been a behavior of dinosaurs as well. If it's a behavior or characteristic seen in only one living archosaur group, but not the other, then the hypothesis might still be right for the dinosaurs, but there isn't as much support for it. If we have evidence for a behavior not seen in either crocodilians or birds, then the hypothesis has no comparative support.

Behavioral Clues in the Rocks

We can start with the evidence from the fossil record. A partial skeleton of a plant-eating dinosaur called *Tenontosaurus* was found in early Cretaceous sediments of Montana. All of the animal's vertebrae were complete and laid out in order from the animal's skull out to the tip of the tail. What was interesting about the skeleton was that its hind legs were separated from its body, and one leg was completely missing, while the other was missing its foot. Both arms were separated from the body as well, and both were missing their hands. The rib cage of the *Tenontosaurus* was disturbed, seemingly pulled open. Most fascinating about the skeleton, however, was that in the sediments around the areas of the legs, arms, and ribs were 11 teeth of the dromaeosaurid dinosaur *Deinonychus*, a close relative of *Velociraptor*. All archosaurs replace their teeth throughout their lives, so it is straightforward to hypothesize that the teeth found around the *Tenontosaurus* were broken from

> *If there is evidence for a particular behavior, and it's a behavior seen in crocodilians and birds, then we can make a pretty good case for it having been a behavior of dinosaurs as well.*
>
> Jack Horner

the mouths of more than one *Deinonychus* as they attacked, killed, and fed upon the carcass.

Desmond Maxwell and John Ostrom, the paleontologists who described this association suggested, on the basis of the number of teeth, that there may have been between six and eight *Deinonychus* involved in the attack.[1] This is the most reliable evidence used to hypothesize that dromaeosaurid dinosaurs hunted in some kind of group similar to pack-hunting predators alive today.

Dinosaur Nesting and Behavior

When it comes to dinosaur nesting and family behaviors, we use the same kind of evidence, logic, and comparisons with modern animals. The Willow Creek Anticline (Fig. 1) in western Montana exposes 50 m (150 ft) of sediment from the 650 m thick, Upper Cretaceous Two Medicine Formation. Within this 50-meter section of strata are four distinct layers that

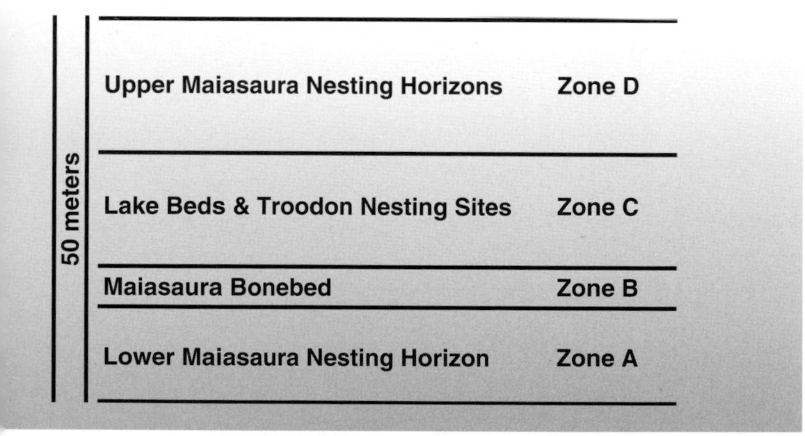

Fig. 1.

Sediment layers yielding behavioral evidence of the Willow Creek Anticline as described in the text.

50 meters		
Upper Maiasaura Nesting Horizons	Zone D
Lake Beds & Troodon Nesting Sites	Zone C
Maiasaura Bonebed	Zone B
Lower Maiasaura Nesting Horizon	Zone A

contain dinosaur fossils, each revealing evidence we can use to hypothesize certain dinosaur social behaviors.

Strata at the base of the Willow Creek Anticline (Zone A of Fig. 1) are represented by green and red mudstones and light gray sandstones. The sandstones are cross-bedded, and this, along with other features, indicates that they were deposited by small streams. The mudstone deposits were formed in the floodplains of these small streams. It is within one layer in this floodplain that we find the dinosaur fossils, two large concentrations of

eggshell fragments, and two distinct concentrations of juvenile dinosaur bones. One concentration consisted of the remains of seven equal-sized baby hadrosaurian (duck-billed) dinosaurs. The second concentration consisted of bones representing 15 individuals, also of equal-sized but larger. Leg bones of the group of seven indicated that each animal was about 45 cm (18 inches) long. Skeletal elements representing the group of 15 indicated these animals were about a meter long, or more than twice that of the group of seven.

When the two sites were excavated, it was noted that the bones of both groups were found in oval-shaped structures. The oval of the structure with 15 individuals measured 2 m (6 ft) in diameter (at the long axis) and 0.5 m deep at its center. The group of seven site was badly eroded, and measured about 1 m across, but was probably at least twice that size before erosion had destroyed one end. Both sites contained small, fossilized twigs, and a few shells of land-snails, but numerous eggshell fragments, some relatively large, were associated with the group of seven. The group of 15 babies was associated with only a few tiny eggshell fragments.

The two oval, bowl-shaped depressions containing baby dinosaur remains have shape characteristics of the nests belonging to a number of crocodilian and bird taxa, and it is therefore hypothesized that these structures are also nests. Studies of the morphology of the skull bones of the babies, compared to an adult skull of *Maiasaura peeblesorum* found in a layer above the nests, indicated that the babies were most likely of the same species.[2] Thus the most likely explanation was that these were nests of *Maiasaura*.

The concentrations of eggshell fragments found within the same layer containing the two groups of babies were discovered to each be confined to areas of about 2 m in diameter, and of similar shape to the bowl-shaped depression containing the 15 young. Because the eggshell concentrations are of similar size to the structures containing babies, and found in the same layer, it is probable that they too represent nest structures, possibly the smashed remains of infertile eggs.

The association of these hypothesized nests within a single layer provides evidence to hypothesize that dinosaurs nested in colonies or groups,[3] much like many modern birds, and to some extent, like some crocodilians.

More Evidence, More Interpretation

Additional data for interpreting nesting behaviors in *Maiasaura* are derived from two mudstone units in the uppermost strata of the Willow Creek Anticline (Zone D, Fig. 1). One site produced a clutch of 16 spherical eggs, each 12 cm (4.7 in) in diameter. A second, badly weathered clutch yielded a few eggs of similar size, and one of the eggs contained embryonic remains, confirming that the eggshells from the eggshell concentrations, and the pieces within the two baby groups, most likely belonged to the same species of dinosaur, hypothesized to be *Maiasaura*.[4] Judging from the size of the eggs, and using a computer to simulate the embryo inside the egg, it was possible to predict the size of the largest baby that could be packed into a spherical egg 12 cm in diameter (Fig. 2). Interestingly, the babies from the group of seven are very close to what is predicted to have been a full-term embryo or hatchling. This probably explains why there was an abundance of eggshells associated with the tiny skeletons.

Maiasaura hatchlings exited their eggs at around half a meter in length, and apparently remained in their respective nests for a period of time in which they at least doubled in length. Histological studies of the ends of the bones of these hatchling and later nestling individuals showed that their leg bones were not ossified, but instead only calcified, and therefore the hatchlings would not have been strong enough for locomotion until they had reached at least 1 m in length.[5,6] For the time it would have taken to double in length, a time that was likely no more than a month, it is hypothesized that the babies were brought food and were protected by a parent. Most crocodilians and birds guard their eggs and their young for some period of time as the young grow, but it is only in some bird groups where parents bring food items directly to the nest to feed the young.

Answering a Question

Over a period of about 20 years, expeditions to the Willow Creek Anticline continued to search the mudstone sediments that contained the *Maiasaura* nests. One goal was to determine the maximum size that the *Maiasaura* babies reached before exiting their nesting areas. The largest bones collected suggest that the babies left the area around the time that they had reached about 1.5 m in length. *Maiasaura* remains representing animals between 1.5 and 3 m in length are extremely rare in any sediments of the Two Medicine Formation.[7] As a result, we are unable to hypothesize about this part of their lives. Between 3 and 8 m in length, however,

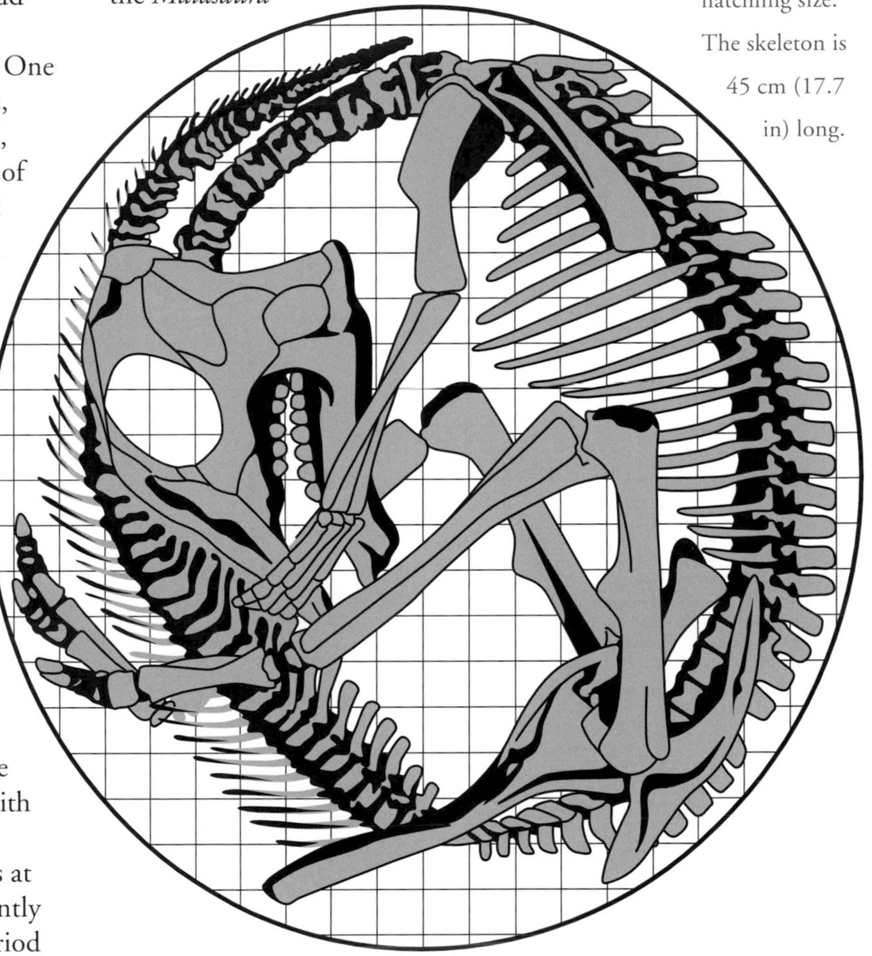

Fig. 2.

Restored full-term hadrosaur embryo packed into a 12 cm (4.7 in) diameter spherical egg to predict hatchling size. The skeleton is 45 cm (17.7 in) long.

there is a sizable amount of data concerning the lives of *Maiasaura*.

Dinosaur Aggregations

Above the red and green mudstones that contain the *Maiasaura* nests at the Willow Creek Anticline is a dark gray mudstone layer (Zone B of Fig. 1) that contains hundreds of thousands of bones, scattered over an area of nearly two square kilometers. It may be the largest dinosaur bone bed on Earth, and it is estimated to contain approximately 10,000 to 15,000 skeletons. The bones are primarily those of hadrosaurs, and all of the hadrosaur bones identifiable to the species level belong to *Maia-*

the Western Interior Seaway located 400 km to the east of the Willow Creek Anticline. Regardless of how the animals died, the fact that they had all apparently died together provides evidence to hypothesize that these animals were living together in some kind of aggregation or herd prior to their deaths.

Additional evidence of potential dinosaur aggregations or herds is provided by other large bone beds found in Cretaceous sediments throughout North American and Asia. Taxa found in large bone beds include the hadrosaurs (duck-bills) *Prosaurolophus, Hypacrosaurus, Gryposaurus, Brachylophosaurus, Edmontosaurus,* and *Shantungosaurus,* and the ceratop-

Fig. 3.

Small section of *Maiasaura* bone bed quarry map showing orientation and association of bones.

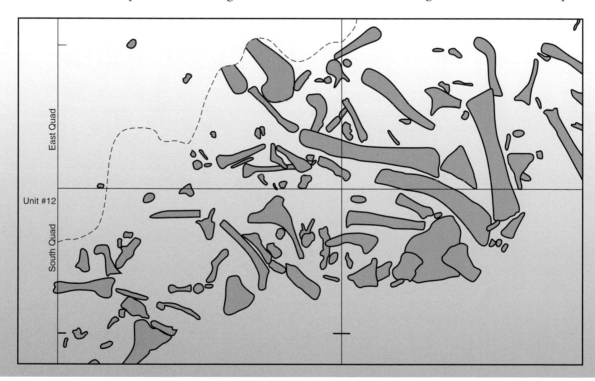

saura peeblesorum. The working hypothesis is that all of the hadrosaur remains are those of *Maiasaura*.

Chemical analyses of the bones from different parts of the bone bed indicate that all the bones were deposited at the same time, rather than over a period of years,[8] suggesting that the animals died simultaneously. Careful study of the bones, and their orientations within the bone bed (Fig. 3) provide evidence that the animals may have died in a violent, catastrophic storm, possibly a hurricane from

sians (horned dinosaurs) *Centrosaurus, Styracosaurus, Einiosaurus,* and *Pachyrhinosaurus.* Many other kinds of dinosaurs are also found in groups, and it can be hypothesized that most, if not all dinosaurs were gregarious, similar to living archosaurs.

Patterns in the strata provide evidence to support the behavioral hypotheses that maiasaurs were social animals, and that they nested in colonies, guarded their eggs and subsequent young, brought food to the growing, nest-bound babies, and eventually traveled in gigan-

tic groups.[9] This group-traveling is occasionally regarded as migratory behavior, although there is no particular evidence for its support.

Evolving Hypotheses

In another layer of the Willow Creek Anticline, above the *Maiasaura* bone bed, is a limestone unit that, based on algal fossils, represents a sequence of shallow alkaline lakes (Zone C of Fig. 1). Along the shores of one lake, and on what appear to have been islands in the lake, are sites containing the fossils of dinosaurs and a variety of other organisms, providing a glimpse of an extinct ecosystem. The interpretation of these sites over the past decade illustrates how the science of paleontology is accomplished, how hypotheses are formed, falsified, and in some cases reformulated. One of the sites thought to have been an island, a site called Egg Mountain, was the first site in the Western Hemisphere to produce clutches of whole eggs attributable to dinosaurs.

When first discovered in 1979, the Egg Mountain site was little more than a grass-covered knoll. But, between the clumps of grass were weathered out chunks of eggs, thousands of eggshell fragments, and hundreds of pieces of little bones. When the bones were picked up and studied, a couple of jaws with teeth revealed that they were consistent with a rare carnivorous dinosaur known as *Troödon*. Although at the time *Troödon* was known only from isolated teeth, the bones were also determined to be consistent with what we thought the bones of *Troödon* should look like. Egg clutches were also discovered at Egg Mountain in 1979, and hypothesized to have been laid by *Troödon* since *Troödon* bones seemed to be common and found in association with the eggs.

In 1980, and during the following three years, the Egg Mountain site was excavated and many new specimens were discovered, including relatively complete skeletons of a new species of small plant-eating dinosaur that would be named *Orodromeus makelai*. Interestingly, *Orodromeus* had bones and teeth very similar to those of *Troödon*, and this discovery seemed to falsify the hypothesis that *Troödon* was the nesting dinosaur at Egg Mountain. The rarity of young *Troödon* bones and the abundance of small *Orodromeus* skeletons and isolated bones seemed to lend more support for the eggs having been derived from *Orodromeus*, because babies often die on nesting grounds of living archosaurs. A new hypothesis was formulated stating that Egg Mountain was the nesting ground for the small plant-eater *Orodromeus*.[10] This hypothesis would hold true for the next 13 years. During that 13-year period a number of important discoveries were made in Zone C of the Willow Creek Anticline (Fig. 1). In 1983 embryonic remains were discovered in a clutch of eggs at a site named Egg Island. Preliminary studies suggested that the embryonic remains were consistent with those of the larger *Orodromeus* skeletons.[11] In 1986 the first *Troödon* skeleton, a small juvenile, was found at Egg Mountain, suggesting that both *Orodromeus* and *Troödon* had lived there.

Ten years after the discovery of embryos at Egg Island, David Varricchio made an important discovery. A hundred kilometers north of Egg Mountain, in sediments of the Two Medicine Formation, Dave discovered the hind end of an adult skeleton of *Troödon*.[12] The skeleton was found on top of what was thought to be a clutch of *Orodromeus* eggs. This discovery would not make sense until 1995 when Mark Norell and others published a paper on a skeleton of *Oviraptor* from Mongolia.[13]

The *Oviraptor* skeleton was found in a squatting position over-top a clutch of eggs, a clutch that had previously been identified as being from the plant-eating dinosaur *Protoceratops*. The eggs had been associated with *Protoceratops* back in 1923 when they were first found, because the skeletons of *Protoceratops* were the most common of dinosaur remains in the sediments producing the eggs. This hypothesis was shown to be false in 1994 when an embryo of *Oviraptor* was found in one of the eggs originally attributed to *Protoceratops*.[14]

Following the publications of the *Oviraptor* embryo and the observation that the squatting *Oviraptor* was very similar to the position of the *Troödon* found in association with a clutch of eggs, the Egg Island embryos were further prepared, and more thoroughly studied. The examination revealed that the embryos were in fact *Troödon*, proving false the

long held hypothesis that *Orodromeus* had been the egg layer, and reconfirming the original hypothesis.

Over the first five years of excavation at Egg Mountain more than a dozen clutches of eggs attributed to *Troödon* were collected from three different nesting horizons (Fig. 4). The eggs were found to have an asymmetrical shape (larger at one end that the other) similar to birds.[9] Most clutches consisted of 12 or 24 eggs arranged in circles with the smaller ends of the eggs pointed downward in the sediment. In effect, the eggs were laid, "planted" half their length into the soil of the nests, leaving their upper ends exposed. Apparently, the thorax of a brooding parent covered the exposed areas of the eggs. In the 1990s another important discovery made at Egg Mountain was a *Troödon* clutch with a distinct sediment rim surrounding the clutch, indicating that *Troödon*

and not in clutch-like circular patterns. These eggs, with eggshell structures similar to some theropods, including birds, are generally found in paired linear rows. An embryo was discovered in one of the mystery eggs, but is too small and immature for identification. One could speculate that two distinct species of dinosaurs nested communally, or that they utilized the nesting ground at different times. This co-occurrence is especially curious since both species appear to have been theropods, and therefore most likely carnivorous.

A Truth for the Moment

Data from Egg Mountain and Egg Island now provide extensive evidence to hypothesize the nesting behaviors of *Troödon* and the paleoecology of its nesting ground. The animals nested in colonies, used the nesting ground on at least

Fig. 4.

West-East cross-section of Egg Mountain showing relative positions of *Troödon* egg clutches. The red star numbered 430 is a skeleton of *Troödon*.

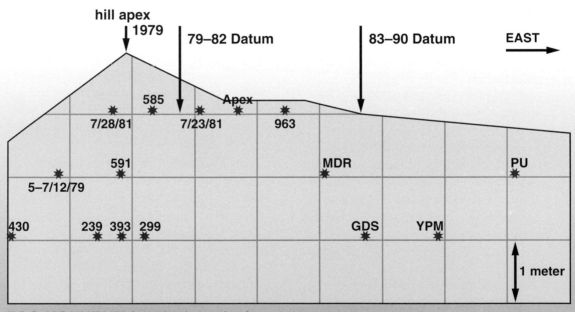

EGG MOUNTAIN (vertical section)
✳ *Troodon* clutches

had deliberately constructed a protective wall around its eggs.[15]

In addition to the egg clutches attributable to *Troödon* on the nesting horizons at Egg Mountain were the eggs of an unidentified dinosaur. The mystery eggs were found lying with their long axes parallel to the soil surfaces,

three different occasions, constructed nests with rimmed borders, arranged their eggs in neat, circular clutches, brooded their eggs by direct body contact, and, apparently brought the carcasses of *Orodromeus* to the nesting area for their hatchlings to feed on. The hatchlings left their respective nests, but may have stayed

in the nesting area for a short period of time before following the adults out of the nesting ground.

Other skeletal remains found on Egg Mountain such as varanid lizards and primitive mammals may represent animals that fed on egg fluids and dead babies.

Over the years since the discoveries at Egg Mountain, many new sites with eggs or young have been found throughout the world, but none has provided comparable behavioral information. New sites in Patagonia, Argentina, have yielded hundreds of nests of sauropod dinosaurs[15] suggesting that these animals also nested in colonies, possibly over consecutive years, but as yet, not providing any evidence of post-hatching behaviors. Many new egg sites have also been found in China and Mongolia, but as many of these sites are in dune sands, there is little evidence of even colonial nesting, although clutch shape is evident. One Mongolian discovery, as yet undescribed, is a group of more than a dozen skeletons of small, juvenile *Protoceratops* arranged in a circular pattern as though in the confines of a nest. This may provide additional evidence to support the hypothesis that some dinosaur young remained in their respective nests for some period following hatching. Some trackways may also provide support for hypotheses concerning dinosaur aggregations.

Dinosaur group and family behaviors are consistent with those of living archosaurs, but as yet there remains precious little data to formulate new hypotheses, or even to test many previous hypotheses. Further discoveries of nesting sites representing other groups of dinosaurs are much needed to help us understand dinosaur social behaviors.

References

1. Maxwell, D., and J.H. Ostrom. 1995. Taphonomy and paleobiological implications of *Tenontosaurus-Deinonychus* associations. *Journal of Vertebrate Paleontology* 15(4):707–712.

2. Horner, J.R., and R. Makela. 1979. Nest of juveniles provides evidence of family structure among dinosaurs. *Nature* 282:296–298.

3. Horner, J.R. 1982. Evidence of colonial nesting and 'site fidelity' among ornithischian dinosaurs. *Nature* 297:675–676.

4. Horner, J.R. 1999. Egg clutches and embryos of two hadrosaurian dinosaurs. *Journal of Vertebrate Paleontology* 19(4):607–611.

5. Horner, J.R., A.J. de Ricqlès, and K. Padian. 2000. Long bone histology of the hadrosaurid dinosaur *Maiasaura peeblesorum*: growth dynamics and physiology based on an ontogenetic series of skeletal elements. *Journal of Vertebrate Paleontology* 20:115–129

6. Horner, J.R., K. Padian, and A.J. de Ricqlès. 2001. Comparative osteohistology of some embryonic and neonatal archosaurs: phylogenetic and behavioral implications for dinosaurs. *Paleobiology* 27:39–58.

7. Horner, J.R. 1994. Comparative taphonomy of some dinosaur and extant bird colonial nesting grounds. In *Dinosaur eggs and babies*, eds. K. Carpenter, K. Hirsch and J.R. Horner, 116–123. Cambridge: Cambridge University Press.

8. Trueman, C.N. 1999. Rare earth element geochemistry and taphonomy of terrestrial vertebrate assemblages. *Palaios* 14:555–568.

9. Horner, J.R. 2000. Dinosaur reproduction and parenting. *Annual Review of Earth and Planetary Sciences* 28:19–45.

10. Horner, J.R. 1987. Ecological and behavioral implications derived from a dinosaur nesting site. In *Dinosaurs past and present*, Vol. 2., eds. S. Czerkas and E.C. Olsen, 50–63. Seattle: University of Washington Press.

11. Horner, J.R., and D.B. Weishampel. 1988. A comparative embryological study of two ornithischian dinosaurs. *Nature* 332:256–257.

12. Varricchio, D.J., F. Jackson, J.J. Borkowski, and J.R. Horner. 1997. Nest and egg clutches of the dinosaur *Troödon formosus* and the evolution of avian reproductive traits. *Nature* 385:247–250.

13. Norell, M.A., J.M. Clark, L.M. Chiappe, and D. Dashzeveg. 1995. A nesting dinosaur. *Nature* 378:774–776.

14. Norell, M.A., J.M. Clark, D. Demberelyin, B. Rhinchen, L.M. Chiappe, A.R. Davidson, M.C. McKenna, P. Altangerel, and M.J. Novacek. 1994. A theropod dinosaur embryo and the affinities of the Flaming Cliffs dinosaur eggs. *Science* 266:779–782.

15. Chiappe, L.M., R.A. Coria, L. Dingus, F. Jackson, A. Chinsamy, and M. Fox. 1998. Sauropod dinosaur embryos from the Late Cretaceous of Patagonia. *Nature* 396:258–261.

Dinosaur Physiology: Hot, Cold, or Lukewarm?

10

Armand de Ricqlès

Laboratoire d'Anatomie Comparée (URA 1137)
Collège de France, Paris

Physiology is the science that tells us how animals and plants work as "living machines." But fossils are dead. How is it then that we gather evidence on dinosaur physiology—about how dinosaurs "worked" when they were alive? The first possibility is to look at living animals, to see what they can do and how they do it, and to make comparisons with dinosaurs. But what do we have to compare?

In living animals, we can check how fast their hearts beat, how they breathe, how long they can run, whether they are warm- or cold-blooded, and many other facets of their lives. Living animals allow us to study both their structures and anatomy (shape, form, size, proportion, structures and relationships of their organs) and their functions (how their organs and muscles work, how they eat, digest, move, reproduce, and behave). We can look at the living animal on many levels: as a whole organism, behaving in its environment, right down to the smallest details of how cells work in organs and how genes work in cells. But as far as dinosaurs go, we have only silent, dry bones and teeth.

In the fossil, all functions have been lost and only some structures are left. Moreover, since what is left is most often only the skeleton, information about the soft parts, the cells, and the genes are generally lost forever.

It is a long way to go from the fragmented skeleton of a dinosaur entombed in rocks to reasonable, scientifically based conclusions about how it once worked as a living organism. What we want to do is to reconstruct the living being of which the fossil is only a distorted and incomplete remnant. How difficult and how reliable is such an exercise?

Let's say we find a reasonably complete skeleton of a fossil horse or crocodile. Even though they may not look exactly like living horses or crocodiles, we have no trouble "seeing" them as living beings. When a direct comparison between a fossil and a living animal is straightforward, common sense easily suggests at once how the fossil "worked" as when alive. We can implicitly

> *. . . because the skeleton is the scaffold of the vertebrate body, it can be used as a surprisingly accurate summary of the complete, living organism, and it can provide a lot of information.*
>
> *Armand de Ricqlès*

Armand de Ricqlès is Professor and Chair of Historical Biology and Evolutionism at the Collège de France in Paris and has held visiting professorships at the University of Chicago and at the Miller Institute for basic research of the University of California, Berkeley. He received his training in biological and geological sciences from the University of Paris and the Paris Museum. He got his Ph.D. in Paleontology from the same in 1973. His general research interests include vertebrate paleobiology and phylogeny at large, as well as their functional morphology and evolution. He has specialized, with his research group in Paris (UMR CNRS 8570), in long-term research on bone as a tissue among extant and extinct vertebrates and has developed the use of bone paleohistology as a tool in Paleobiology.

conclude, for instance, that the fossil horse was a warm-blooded, galloping plant-eater and that the fossil crocodile was a sprawling, amphibious, cold-blooded flesh-eater. To reach these conclusions, we rely—even if unconsciously— on principles such as, "like causes produce like effects" and "there is a necessary link between a given shape or structure and its purpose or function," so "similar structures will support similar functions."[1, 2]

But nothing is alive that really approximates a dinosaur (or is there?), and so no easy, direct comparisons to dinosaurs can be drawn from living animals. In order to reconstruct a "living dinosaur" we have to rely on more complex and indirect lines of evidence than we used to compare fossil horses or crocodiles to living ones. Also, we may be less sure about our evidence and will need several lines of evidence that are independent from each other, in order to reach a common conclusion that we hope sounds reasonable.

Experimental sciences use demonstrative evidence (from experiments), whereas historical sciences mostly use accumulative (or inductive) evidence. Accumulative evidence refers to a situation when many of independent observations and lines of evidence converge on a common explanation or conclusion. Each line alone offers only a weak inference, but their agreement creates a bundle of presumptive support that may ultimately foster a given conclusion beyond reasonable doubt.[3]

We will never devise experiments that demonstrate the life and fate of Alexander the Great, but we have a copious body of indirect facts and data that show (not prove!), beyond any reasonable doubt, that he was a real human being. In much the same way, we can devise complex, reliable, and testable reconstructions of many aspects of the "lives" of ancient organisms.

Examining Physiology

Any vertebrate animal, whether a fish, a bird, or a mammal, is a highly complex organism formed by well over 250 different types of specialized cells. Physiology deals with all the conditions and mechanisms that allow an organism's cells to live together and function

properly. Cell types are organized into organs, which interact so that an organism as a whole "works" and maintains its integrity in the face of its environment. Physiology does not describe organs for their own sake (this is the task of anatomy) but rather analyzes and deciphers their roles and functions, and how they interact and integrate as regulated systems.

Thus, physiology is not a science of structures, but a science of functions or processes. It captures the dynamics of life itself. The science of physiology is based on quantification of data from living organisms, including blood pressure, amount of sugar in the blood, amount of oxygen consumed per hour, or whatever other measurements prove relevant. The mechanisms that generate these measurements are understood by experiments on living organisms, organs, or cells.

The part of physiology that deals with heat production and regulation of body temperature and levels of energy is referred to as thermometabolism.[4] Metabolism may be roughly understood as the processes and rates by which food is converted by an individual into its own living matter, available energy, and wastes. An important by-product of metabolism is heat, an inescapable result of chemical reactions. This heat can be used by the body, not just wasted. Animals follow different "strategies" to optimize the ways their metabolisms cope with the challenges of their environment. We are interested in what dinosaurs did in that regard, given what we can learn from their bones: were they warm-blooded? Cold-blooded? Or something in-between?

Physiological Information from Vertebrate Fossils

Fossils are mostly skeletons and nothing more. However, because the skeleton is the scaffold of the vertebrate body, it can be used as a surprisingly accurate summary of the complete, living organism, and it can provide a lot of information. This information pertains directly to the body structures or anatomy, and only indirectly to physiology, nevertheless it allows precise conclusions about some body functions. During the early 19th century, Cuvier used

comparative anatomy to propose the principle "functional relationships" among the various parts of the body, which allow it to work as a whole to fit the demands of life.[5] For example, the shape and structure of the teeth of a flesh-eating mammal are quite different from those of a plant-eater, just as their sensory organs, brains, guts, jaws, limbs, and horny appendages also differ. The various parts of the body are necessarily coordinated with their functional demands. One would not expect hooves on a tiger and claws on a cow! If all the various structural characters of an organism, taken together, tell us how the organism works, then even a few isolated characters, such as the shape of a tooth, may suggest the characters and functions of the other missing parts. Let's go back to our fossil crocodiles and horses. From certain diagnostic features such as their lower jaws, we can speculate on the reconstruction of their entire skeletons, and from there, on their way of life, with a reasonable degree of likelihood.[2]

Given this general principle, paleontologists can reconstruct with reasonable accuracy some behaviors and traits of extinct animals, even if they're different from living ones. Teeth and jaws can tell us a lot about feeding. Limb postures, structures, joints, ratios of various limb bone lengths, etc. allow us to interpret how animals moved. How much an animal depended on a given sense (sight, hearing, smell) over another can be reconstructed by studying the relative size and shape of various skull regions, as well as details of the brain cavity, which provide an indirect "cast" of the brain shape and size itself. In addition, the adaptation of an extinct animal to a general ecological role (aquatic, amphibious, terrestrial, arboreal, airborne) can generally be deciphered.

Warm *versus* Cold Blood: Is It Really the Right Question?

The separation of animals into "cold-blooded" and "warm-blooded" is at best a gross simplification.[4] Among living animals, it is well known that some moths reach a high body temperature during flight and cannot take off in the morning before "warming up their engines" by shivering. Insects are nevertheless regarded as "cold-blooded." Some fishes, called "cold-blooded," are specialized to live in hot springs, where their body temperature is constantly higher than those of birds! Large monitor lizards like the Komodo dragon, called "cold-blooded" reptiles, are active foragers whose muscular activity warms them to temperatures higher than the surrounding air, and their bodies are large enough to keep them warm even at night. And, small lizards get so much heat from the sun that their body temperatures are as high as those of mammals, allowing them great activity and speed.

Obviously, from such examples the "cold- vs. warm-blooded" dichotomy is too simplistic to explain the thermal physiology of living animals. Even a thermometer is not helpful, because the body temperature of an animal tells us very little about its general metabolism and physiology! If the situation is already so complex for living species, the question is even more tantalizing for fossils. When we ask whether dinosaurs were "cold- or warm-blooded," what do we want really to know?

Probably what we have in mind is whether dinosaurs had lifestyles roughly similar to large living reptiles like turtles, monitor lizards, or crocodiles, or whether they were more like large ground birds (ostriches), or even large mammals.[6, 3, 7] Rather than retaining the crude division between "cold- and warm-blood," we need more precise intellectual tools. We should first distinguish between (a) the source of an animal's heat, and (b) the constancy and regulation of body temperature.

a) Source of the heat: Animals can be roughly divided into two groups: endotherms and ectotherms. Basically, this grouping deals with the source of metabolic heat. In order to be endothermic, you have to be able to burn food at a high rate, producing heat from within, hence the word endotherms (endo= from within). Mammals and birds are typical endotherms. Conversely, most animals have a much "lazier" metabolism (five to ten times slower than in typical endotherms) and do not produce significant amounts of heat. They have to rely on external sources of heat and are called ectotherms (ecto= from without).

b) Constancy and regulation of an animal's body temperature is quite another matter. Some animals can regulate their body temperature at a given value at which their enzymes and other body systems perform best. Their physiology allows them to produce more heat, e.g., by shivering, in cold environments and to dissipate more heat, e.g., by perspiration, in hot environments, so they can maintain a roughly constant core temperature. Such animals, called homeotherms (homeo= same), are typically living mammals and birds. Conversely, all other living animals generally have variable body temperatures, more or less consistent with their environments, with no specialized physiological system of temperature regulation. Such animals are called poikilotherms.

However, the metabolic strategies of poikilotherms are highly variable. Some poikilotherms can function comfortably under highly varied temperatures, whereas others can only be active in a narrow temperature range. Some also have limited behavioral control of their own body temperatures, and thus reach some practical level of homeothermy. Such behavior is common to lizards that change microhabitats during their daily cycles, alternatively basking in the sun and cooling in the shade to maintain a roughly constant, optimal core temperature. Finally, some poikilotherms can exercise their muscles to generate significant amounts of heat, raising their core temperature above that of their immediate surroundings.

From this review, it is clear that among living animals, the exception is the rule. However, the general condition of mammals and birds is both endothermy and homeothermy, i.e., the typical warm-bloodedness that is supported by an active metabolism.[4]

Locomotion and metabolism also seem to be linked. Birds and mammals rely mostly on aerobic metabolism for their locomotion. They use oxygen to "burn their fuel" efficiently and to sustain protracted exercise. In contrast, living reptiles can run just as fast as mammals, but in order to do so they must rely on anaerobic metabolism. "Burning their fuel" without oxygen soon produces torpor, which constrains how much they can do. Therefore they are commonly "sit and wait" ambush hunters.

Body size is also important. The relative "cost" of being an endotherm decreases as body size increases, simply because the ratio of body surface to body volume decreases with larger sizes. In the largest dinosaurs, the thermal inertia of the body (the tendency to retain heat) was such that the animal could have been "warm-blooded" even with a relatively low metabolism. In such cases, endo- and ectothermic physiologies converge to the point that the "warm-blooded" question becomes meaningless. The main problem is heat dissipation.[6, 7, 8, 9, 10, 11]

An Evolutionary Perspective

From an evolutionary point of view, it is clear that "cold-bloodedness" (ectothermy and poikilothermy) is the "generalized" physiological condition. Among vertebrates, this condition is retained among most fishes, amphibians, and reptiles. Thus, ecto- and poikilothermy cover a lot of different situations. Conversely, the "warm-bloodedness" of living birds and mammals (endothermic homeothermy) is a more "specialized," or "evolved" condition among vertebrates, and accordingly covers more limited physiological conditions. Obviously, the endothermic homeothermy of birds and mammals had to evolve from more "generalized" physiological conditions found in their respective ancestors. This change was not a sudden switch from an ectothermic, poikilothermic ancestor to a fully-fledged, endothermic, homeothermic offspring. Rather it was a stepwise evolutionary process.[12, 13]

We use the terms "generalized" or "specialized" to emphasize the idea that "warm-bloodedness" by itself does not represent a "higher" or "better" condition than "cold-bloodedness." Warm-bloodedness is merely another strategy to cope with the business of living under the constraints of a given environment.[11] Each strategy has advantages and disadvantages. Endotherms can do things, such as maintain active lifestyles in conditions of severe cold, that ectotherms apparently cannot do. But endothermy is an expensive lifestyle. With small to medium body sizes, endotherms are "gas guzzlers," placing a high demand on food resources. They need about 10 times

more food than ectotherms of the same body mass just to produce enough heat to keep their body at a high, constant temperature. Ectothermy has its own advantages in the struggle for life, such as its relatively cheap demand on environmental resources to sustain life, and accordingly its greater efficiency in turning a given amount of food into more living matter (biomass) than endotherms can do. After all, there are more living species of ectothermic vertebrates than of endotherms! We must keep in mind all these various concepts and data when discussing the possible thermometabolic situations in dinosaurs.

Paleothermometers: Three Approaches to Identify Metabolic Strategies of Dinosaurs

1) Ecological approach Let's consider a given terrestrial environment that forms a large, closed ecosystem. The vegetation in this environment should sustain the herbivorous vertebrates that live there. In turn, the herbivores should sustain a smaller number of carnivorous vertebrates that prey upon them. This scenario is the familiar idea of the "food pyramid" with predators at the top. Since ectotherms don't need to consume food just to build up body heat, they are much more sparing of their environment than endotherms. Therefore, a given environment should be able to sustain about 10 times more ectotherms (or a larger biomass) than endotherms. Similarly, a given biomass of herbivorous vertebrates should sustain a much larger biomass of ectothermic than endothermic predators. Thus, within any ecological community of vertebrates, the biomasses will differ depending on the number of endotherms and ectotherms present.[6]

This hypothesis can be tested in living communities dominated by large endothermic mammals, such as African savannas, where we can determine the actual values of biomass of herbivorous and carnivorous mammals per square kilometer. How does this help us learn about dinosaurs? If dinosaur communities were composed of endotherms, we'd expect to find values roughly similar to those found in the savanna. Conversely, if dinosaur communities

were composed of ectotherms, the biomass values would be much higher. Early tests of this approach by a census of contemporary herbivorous and carnivorous dinosaur carcasses for a given stratigraphic layer strongly suggested a high prey-predator ratio, and thus implied "endothermic" community values for the dinosaurs. However, many critics questioned the methods used. Some pointed out that, for various practical reasons, the census of dinosaur remains was not unbiased, and so could not provide reliable estimates of the actual populations. Other studies offered results that suggested "ectothermic" community values for dinosaurs.[9, 10] Thus, the ecological approach remains inconclusive, mostly because so many practical difficulties are involved in taking the census of a fossil community.

2) Studying Behavior We can indirectly reconstruct some dinosaur behaviors from their nesting sites, gut and dung contents, and other fossil evidence. If dinosaurs behaved in ways that are uncommon or unknown among living ectotherms, but that may be more readily observed among living endotherms, then such behaviors would at least suggest that dinosaurs employed physiological strategies somewhat more comparable to those of living endotherms than to ectotherms.

The study of dinosaur nesting sites has revealed interesting behaviors: (1) a return to nesting grounds over generations, as in birds, with nesting sites perhaps occupied over several millennia; and (2) strong evidence that at least some dinosaurs evolved parental care.[14, 15] This behavior has been deduced from the discovery of several nests of a species of duckbill dinosaur, *Maiasaura*, in Montana. Some nests of this species contain hatchling young, about 30 cm long, associated with reasonably complete eggshells. Other nests contain only a few broken eggshells and remains of larger young, about 90 to 110 cm long. This finding strongly suggests that the young stayed in the nest for quite some time after hatching, and since they grew so much they had to be fed by adults. Again, this observation in itself does not suggest endothermy, but is simply accordant with the possibility of bird-like behaviors among dinosaurs.

More compelling behavioral evidence for dinosaur endothermy is provided by *Oviraptor*, a small, strange-looking theropod. When first discovered, it was thought that this dinosaur was an "egg robber," hence its name. But recent, detailed studies now suggest that the animal was simply brooding over its own nest of eggs![16]

3) Geography During the Mesozoic, dinosaurs roamed each and every continent, and even temporarily colonized small islands that emerged from time to time from the epicontinental seas. A case in point is their occurrence on lands at high latitudes, such as the northern slope of Alaska, Antarctica, and the southern coasts of Australia.[17] Taking into account the geographic position of those regions during the Mesozoic, it appears that there were indeed polar dinosaurs. Interestingly, birds and mammals (endotherms) are also known from some of those same fossil beds; but generally no turtles, lizards, snakes, or amphibians (all known to be ectotherms) have been found with the polar dinosaurs. This finding strongly suggests that those dinosaurs were able to cope with harsh climatic conditions, along with animals known to be endotherms, whereas those ecosystems were generally devoid of animals known to be ectotherms.[10] Again, these data suggests that the dinosaurs themselves probably possessed some form of endothermic physiology that allowed them to exploit such ecosystems.[13] Of course, climates were milder during the Mesozoic, and conditions at high latitudes were probably far less stringent than they are now. Nevertheless, the extremely wide latitudinal distribution of dinosaurs during the Mesozoic, compared to those of known ectotherms, remains a strong, if not compelling, argument favoring some kind of endothermic physiology.

Anatomy as an Additional Paleothermometer

Upright posture In contrast to most living and fossil amphibians and reptiles, almost all non-avian dinosaurs, both bipedal or quadrupedal, had an upright stance with legs under the pelvis—not out to the side. This posture, shared with most mammals and with all birds, and common to all dinosaurs, was already rec-

ognized by Richard Owen in 1842 when he conceived dinosaurs as a distinct group of advanced reptile. In the late 1960s, emphasis was put on the possible correlation between upright posture and endothermy. Mammals and birds are the only living endothermic tetrapods. They are also the only ones to have an erect stance, contrasting with the sprawling posture of living ectothermic reptiles and amphibians refined.[18] The correlation is not perfect however, since some mammals, such as the platypus, are "sprawlers." The relevant point may be that no living ectothermic amphibians or reptiles have such a sustained upright posture. Accordingly, a case was devised to link endothermy with an upright posture, which would thus suggest endothermy for dinosaurs. The argument has been discussed at length and further refined.[18, 6, 11, 13] At the very least, and set in its proper evolutionary context, it offers strong circumstantial evidence in favor of "specialized" locomotory behavior in dinosaurs, with reliance on commensurate aerobic metabolism.

Nasal turbinates The presence or absence of nasal turbinates in dinosaurs offers additional anatomical evidence about temperature regulation. Turbinates are thin, curled blades of bone or cartilage that support soft tissues in the air passage behind the nostrils in mammals and birds (Fig. 1). These structures have various functions, including control of temperature and humidity of the respiratory air going into and out of the lungs. In the 1990s, some scientists felt that the occurrence of turbinates was the "magic bullet" that separated endotherms from ectotherms, because living endotherms have turbinates and ectotherms do not.[10] Now, we have to be careful with our reasoning here. If good evidence of extensive turbinates was ever demonstrated in a fossil, it would then offer support for endothermy.[19] However, lack of turbinates would be just that: negative evidence, demonstrating nothing. Computed Tomography (CT) scanning of the nasal passages of various dinosaurs suggested no turbinates. Some claimed this lack as evidence that dinosaurs were ectothermic. However, as noted, very little can be made of this negative evidence. Some living birds have

highly reduced turbinates or none at all, and cartilaginous turbinates, even if present during life, would not be preserved in fossils. A correlation between occurrence of turbinates and endothermy is not absolute, and the real functional significance of those structures is not clear either.

Blood pressure The erect posture and large sizes of dinosaurs imply very great vertical distances (and hence blood pressure levels) between the feet, heart, and brain. The heartbeats would have to generate high pressure in order to reach the brain and other organs sufficiently. This high pressure would be possible only if the right and left parts of the heart were completely divided, thus separating the lesser (lungs/heart) and greater (body/heart) blood circulations. While all extant ectothermic vertebrates mix the two circulations to some degree, the complete separation of circulations is precisely the situation seen in all living endothermic vertebrates, mammals and birds. The physical constraints of many dinosaurs strongly suggest an endothermic-like circulatory system. This observation does not by itself demonstrate endothermy, but again strongly suggests its possibility in dinosaurs. [3, 9, 10]

Skin/skin appendages Another obvious anatomical distinction between living endo- and ectotherms is the structure of the skin. Endotherms have so-called phaneras (phaneros = what can be seen), namely fur or hair in mammals and feathers in birds. These features are highly specialized horny productions of the skin that actually protrude to create the outer shape of the animal, as seen with its fur or plumage. Phaneras have many functions, including signaling or camouflage, sensory functions, aerodynamics (birds' wings), and obviously, thermal insulation, an especially vital function in small endotherms, as already stressed. Conversely, ectotherms lack typical phaneras. The skin of amphibians is naked and reptiles have horny scales that closely outline the actual body surface. In both cases, the color and shape of the animal is defined by the skin itself. Instead of thermal insulation, the skin of ectotherms allows extensive heat exchange with the environment.

Fossil evidence for skin and skin appendages is scarce, but of crucial value. Evidence of feathers in early fossil birds (such as *Archaeopteryx* and *Confuciusornis*) strongly suggests that they were already endothermic. Even Richard Owen noted this in 1870! Similar discoveries of "fur" in small, early mammals point to the same conclusion.[20] What about dinosaurs? Skin impressions from large dinosaurs suggest a naked, pebbly skin, somewhat like the covering

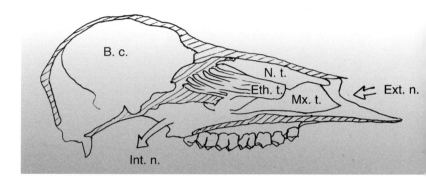

of a golf ball. This does not necessarily suggest ectothermy, because large endotherms such as elephants also lack phaneras. Discoveries of tiny dinosaurs in China, on the other hand, such as *Beipiaosaurus*, *Sinosauropteryx*, *Protarchaeopteryx*, and *Caudipteryx*, have clearly preserved phaneras of various shapes, from hairlike fringes to fully vaned and barbed feathers.[21,20] Their presence strongly suggests insulatory functions and very likely endothermy of those small dinosaurs.

Paleothermometers: Phylogeny

Among dinosaurs, theropods share with birds a high number of specialized anatomical characters that indicate common ancestry.[11,7] Other groups of dinosaurs, such as the giant sauropods and the duckbill, horned, and armored dinosaurs, are, according to their anatomies, more and more remote relatives. This general structure of kinship among birds, dinos, and finally crocodiles (Fig. 2) strongly suggests that the evolutionary changes in physiology culminating in avian endothermy had already taken place in dinosaurs and perhaps even among some of their more "primitive" archosaurian relatives. If so, it would be logical to think that all dinosaurs already shared at least some of the

Fig. 1.

Slightly off-centered and simplified longitudinal section of a deer skull. Turbinates are thin curled bone or cartilage blades developed in the nasal passage. B.c., Brain cavity; Ext. n., External nostrils; Int. n., Internal nostrils or choanae; Eth. t., Mx. t., and N.t., turbinate bone/cartilage.

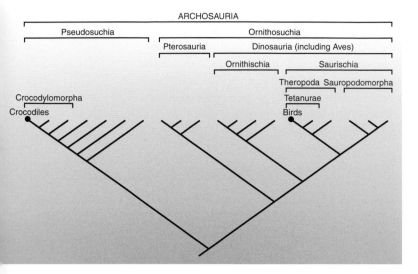

ARCHOSAURIA

Pseudosuchia | Ornithosuchia

Pterosauria | Dinosauria (including Aves)

Ornithischia | Saurischia

Theropoda Sauropodomorpha

Tetanurae

Birds

Crocodylomorpha
Crocodiles

Fig. 2.

A simplified cladogram of archosaurs showing evolutionary relationships.

physiological advances leading to "true" bird-like endothermy, and that the more they shared, the closer they are to birds. Again, in that context, the discovery of "feather-like" phaneras in small theropods fits this hypothesis very well.[21, 13]

Paleothermometers: Bone Tissue Structures and Growth Strategies

In the late 1960s, a new approach to deciphering the physiology of dinosaurs was put forward—the study of their bone microstructures, or paleohistology. Histology is the science that studies and describes tissues at a microscopic level of detail.[22] How is it possible to study these tissues in fossils?

All organs of the body, and bones are no exception, are formed of cells and cell-specific products known as extracellular matrices. Cells and their matrices are organized in highly specific ways to form tissues. Skeletal tissues are hard because they become extensively mineralized during life. Because skeletal tissues are already mineralized, they resist decay after death and can become fossilized later, if circumstances allow. In the process, they can retain all the minute microscopic details of their living form. Hence, it is possible to study the fine structure of dinosaur bone and teeth almost as precisely as those of living vertebrates (Fig. 3).[23, 15]

Bone histology can tell us a lot about the biology of living vertebrates. Bone preserves the record of its growth in its structure while it

is laid down in the growing animal. It is possible to decipher the message of bone recorded in its fine structure in living animals, where life history data such as growth, longevity, ecology, and physiology are readily available by observation and experiment.[24]

Among living vertebrates, there is a general relationship among thermometabolic physiology, growth patterns, and bone histology. Actually, there is a necessary functional connection, and bone histology works as a recorder of the growth patterns, hence the underlying metabolic strategies. In living animals, there is generally a sharp physiological contrast between endo-homeothermic birds and mammals and ecto-poikilothermic amphibians and reptiles. This contrast can be observed readily in their respective bone structures. Large mammals and birds, for instance, routinely have bone tissues that are specialized for fast growth and intensive bone turnover throughout life. In contrast, large living reptiles generally have slow-growing bone tissues and slower bone turnover. Such bone is routinely interrupted by a number of "lines of arrested growth." These cyclical growth marks have been experimentally demonstrated to be annual in many cases, and they often reflect the interruption of growth of ectothermic animals during the yearly "bad" season (estivation, hibernation, drought, heat, cold, or low food supply).[23, 24]

Given such knowledge from living animals, it may be possible to analyze the data recorded in fossil bone and interpret the biology of the fossils. Of course, one has to assume that similar causes will produce similar effects, and that the processes that we see today also operated in the past. Early students of dinosaur bone tissues were struck by the great similarity of dinosaur bone tissues to those of large mammals and ground birds, rather than to those of living amphibians and reptiles.[18, 3] Later studies confirmed that such features in all likelihood point toward high to very high growth rates in dinosaurs, allowing most of them to reach their large body sizes surprisingly quickly, at rates unknown in ectothermic vertebrates.[10, 11, 12]

These data imply a physiological situation for dinosaurs closer to those of mammals and birds than to those of living reptiles, because a

high level of metabolism seems necessary to sustain such active growth. On the other hand, dinosaur bone is not typically "mammalian." It generally shows more conspicuous growth cycles than do most mammalian bones. A minimal assessment would be to grant dinosaurs a somewhat "intermediate" physiology, perhaps not geared fully to mammalian or bird-like levels, but at some intermediate (and possibly varied) levels between typical living reptilian and mammalian ones.[8,10] This conservative version of dinosaur physiology is supported by the extensive occurrence of growth rings in dinosaur bones, which may suggest a less than fully endo-homeothermic status. However, some dinos don't have any growth rings in their bones.[7]

Interestingly, two other groups of fossil vertebrates share with dinosaurs bone tissue characters that suggest a somewhat elevated metabolism, compared to living reptiles.[18] One is the pterosaurs, the flying reptiles of the Mesozoic, widely regarded as endothermic because of the metabolic requirements of flight. The other one is the so-called therapsids, which form the ancestral group from which the endothermic mammals evolved. Most other groups of fossil amphibians and reptiles have bone tissue histologies akin to those of large ectotherms such as turtles and crocodiles.[23, 24, 12]

Altogether, those data offer a general picture of the evolution of "warm-bloodedness" among vertebrates, suggesting that endothermy evolved at least twice independently—once in the lineage leading to dinosaurs, pterosaurs and birds, and once in the mammalian lineage.

Conclusion

Have we demonstrated that dinosaurs were warm- or cold-blooded? No. What we have done is to show that the question itself—as stated in such terms—is clouded by ambiguities. What we had to do, then, was to rephrase the question more precisely, to raise more precise issues that can be addressed by scientific methods.

Among the various "paleothermometers" we have tried to devise and use to get an idea of dinosaur physiology, it is obvious that none

is absolutely reliable and offers a definitive, absolute answer. The answers are all ambiguous to varying degrees, depending on the "thermometer" used. However, the various "thermometers" all offer strong suggestions that dinosaur physiology was neither like that found in living fishes, amphibians, and reptiles (which are assumed to represent the "generalized" physiological condition), nor exactly like that found in birds and mammals (which are

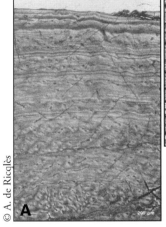
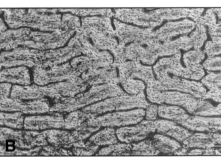

© A. de Ricqlès

Fig. 3. Bone histology of two archosaurs. **A.** Detail of bone tissue from a femur shaft of *Coelophysis*, an Upper Triassic theropod dinosaur. Well-vascularized bone tissue is indicative of fast growth and can already be seen in this small, early dinosaur. This type of bone tissue is also found in most other dinosaurs and in large extant birds and mammals. **B.** Bone shaft cross-section, at low magnification, of a Triassic pseudosuchian closely related to crocodilians. This less vascularized bone tissue type is linked to lower growth rates and is seen in both extant and extinct ectothermic vertebrates.

assumed to represent the more "evolved" physiological condition among vertebrates).

It is not surprising that the answer we get is neither one extreme nor the other; it is instead a compromise. The contrast between endo- and ectothermy is not a sharp dichotomy. There is no "either, or" but a continuum of possible situations. If we accept that birds evolved from and are a kind of theropod dinosaur, birds do "nest" evolutionarily within dinosaurs as a group (Fig. 2), and it is all the more likely that evolution toward avian "warm-bloodedness" took place step by step within the dinosaurs and even before.[21, 12] Given the most recent data, it is thus likely that most dinosaurs shared a more "special-

ized" thermometabolic physiology than living reptiles do, but it is unlikely that they used exactly the same strategies as large ground birds or large mammals today. As more lines of evidence ("paleothermometers") are explored in the future, we may come closer to understanding more precisely how dinosaurs built up and regulated their body temperatures.

References

1. Schultze, H.P., and L. Trueb (eds.). 1991. *Origins of the higher groups of tetrapods: controversy and consensus*. Ithaca, NY: Comstock Publishing Associates/Cornell University Press.

2. Thomasson, J.J. (ed.). 1995. *Functional morphology in vertebrate paleontology*. New York: Cambridge University Press.

3. Thomas, R.D.K., and E.C. Olson (eds.). 1980. A cold look at the warm-blooded dinosaurs. AAAS Symposium 28. Boulder, CO: Westview Press.

4. Whittow, G.C. (ed.). 1970-73. *Comparative physiology of thermoregulation*, Vols. 1–3. New York and London: Academic Press.

5. Appel, T.A. 1987. *The Cuvier-Geoffroy debate: French biology in the decades before Darwin*. New York: Oxford University Press.

6. Bakker, R.T. 1986. *The dinosaur heresies*. New York: William Morrow and Co., Inc.

7. Paul, G.S. (ed.). 2000. *The Scientific American book of dinosaurs*. New York: St Martin's Press.

8. Dunham, A.E., K.L. Overall, W.P. Porter, and C.A. Forster. 1989. Implications of ecological energetics and biophysical and developmental constraints for life-history variations in dinosaurs. In *Paleobiology of the dinosaurs*, ed. J.O. Farlow. Geological Society of America Special Paper 238. Boulder, CO.

9. Weishampel, D.B., P. Dodson, and H. Olsmolska (eds.). 1990. *The dinosauria*. Berkeley: University of California Press.

10. Farlow, J.O., and M.K. Brett-Surman (eds.). 1997. *The complete dinosaur*. Bloomington: University of Indiana Press.

11. Currie, P.J., and K. Padian (eds.). 1997. *The encyclopedia of dinosaurs*. San Diego: Academic Press.

12. Padian, K., A. de Ricqlès, and J.R. Horner. 2001. Dinosaurian growth rates and bird origins. *Nature* 412:405–408.

13. Schweitzer, M.H., and C.L. Marschall. 2001. A molecular model for the evolution of endothermy in the theropod-bird lineage. *Journal of Experimental Zoology* 291:317–338.

14. Carpenter, K., K.F. Hirsch, and J.R. Horner (eds.). 1994. *Dinosaur eggs and babies*. New York: Cambridge University Press.

15. Horner, J.R., and J. Gorman. 1995. *Digging dinosaurs*. New York: Harper Collins.

16. Norell, M.A., J.M. Clark, L.M. Chiappe, and D. Dashzeveg. 1995. A nesting dinosaur. *Nature* 378:774–776.

17. Ostrom, J.H. 1970. Terrestrial vertebrates as indicators of Mesozoic climates, 347-376. In *North American Paleontology Convention Proceedings D*.

18. Bellairs, A. d'A., and B. Cox (eds.). 1976. Morphology and biology of reptiles. *Linnaean Society Symposium Series*, Number 3. London: Academic Press.

19. Hillenius, W.J. 1994. The turbinates of therapsids: evidence for late Permian origins of mammalian endothermy. *Evolution* 48:207–229.

20. Gee, H. (ed.). 2001. *Rise of the dragon*. Chicago: University of Chicago Press.

21. Gauthier, J., and L.F. Gall (eds.). 2001. *New perspectives on the origin and early evolution of birds: proceedings of the international symposium in honor of John H. Ostrom*. Peabody Museum of Natural History, Yale University, New Haven, CT.

22. Cormack, D. 1987. *Ham's histology*. New York: Lippincott.

23. Carter, J.G. (ed.). 1990. *Skeletal biomineralization: patterns, processes and evolutionary trends*, Vol 1. New York: Van Nostrand-Reinhold.

24. Hall, B.K. (ed.). 1991-93. *Bone*, Vols. 1–7. Boca Raton, FL: CRC Press.

Feathers and Flight: Current Ideas

11

Philip J. Currie
Royal Tyrrell Museum of Palaeontology
Drumheller, Alberta, Canada

People have a fascination with other animals, although different people seem to be attracted to different kinds. Some people love dogs; some prefer cats; some horses; some are completely fascinated by whales, or snakes, or aquarium fish. The list of preferences is as extensive as our tastes in music or food. But if we took a vote on what people prefer to know more about, two of the most popular groups would be birds and dinosaurs. Birds are not only a constant source of interest to us, but they are also one of the most successful groups of animals with backbones. Over 10,000 species of birds are alive today. In a sense, that makes them more than twice as successful as mammals, of which there are only 4,000 living species. We love to watch birds in the wild and in zoos, and to keep them as pets. Another group of animals that we have a total fascination with is known as the Dinosauria. To most people, dinosaurs and birds are as different as day and night. That is why many people are shocked when they hear modern biologists and paleontologists state that birds are in fact living dinosaurs.

Technically, under a modern biological classification, birds are part of the Dinosauria. You can correctly call them dinosaurs if you like, although it is also correct to call them birds. All birds are dinosaurs, but not all dinosaurs are birds. Think of it the same way as you do cats and dogs; all cats and dogs are mammals, but not all mammals are cats and dogs. We can refer to our pets as mammals, but we are more likely to call them cats or dogs. There is no problem with referring to our modern feathered friends as "birds" or "avians." However, technically we should say "non-avian dinosaurs" or "non-avian theropods" if we want to refer to all dinosaurs or theropods (non-flying, meat-eating dinosaurs) other than birds. Please remember that when I say "dinosaur" or "theropod" in this article, I am using these words informally to mean "non-avian dinosaur" and "non-avian theropod". However, when I use the technical terms "Dinosauria" and "Theropoda," birds are included.

> *. . . the overwhelming evidence shows us that birds evolved directly from theropod dinosaurs. No other candidates exist with so many anatomical similarities . . .*
>
> *Philip Currie*

Philip J. Currie is the Curator of Dinosaurs at the Royal Tyrrell Museum of Palaeontology, and an adjunct professor at the University of Calgary. He received his B.S. at the University of Toronto, and M.S. and Ph.D. from McGill University. For the past 25 years, he has been collecting dinosaurs in Alberta. He was co-leader of the Canada-China Dinosaur Project (1986–1991), the Argentina-Canada Dinosaur Project (1997 to present), and Nomadic Expedition's Dinosaurs of the Gobi (1996 to present). In addition to his research papers on theropods and other dinosaurs, he has done much to popularize dinosaurs through his books, magazine articles, lectures, and media interviews.

Changing Ideas of Bird Origins

Bird fossils are almost as rare as the scientists who study them, because most birds are small and have fragile, hollow bones. When they die they are far more likely to be eaten or destroyed than they are to be fossilized. The popular concept of dinosaurs, on the other hand, suggests that most were enormous animals—including the largest to ever walk the Earth. Huge animals have huge bones, which are less likely to be eaten or destroyed by physical factors or bacteria, and are therefore more likely to be buried and fossilized. Our concept of the relationships of birds and dinosaurs has changed in recent years, partly because of improved techniques of analysis, but largely because of the discovery of more and better preserved fossils of birds and small dinosaurs.

Archaeopteryx (Fig.1) is generally considered to be the first (or earliest) bird.[1] Originally discovered in 1861 in Solnhofen, Germany, *Archaeopteryx'* remarkably well-preserved fossils still provide us with some of the best informa-

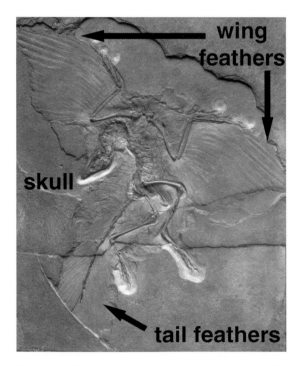

Fig. 1. The Berlin *Archaeopteryx* clearly shows a mix of dinosaur and bird characters. There are long feathers on the arms, but the fingers are clawed and separate like those of a carnivorous dinosaur. The long tail feathers are bird-like, although no bird today has such a long bony tail.

tion on the ancestry of birds. This small, chicken-sized animal was immediately recognized as a bird, but only because feathers were preserved in association with the skeleton. If the feathers had not been there, it would have been identified as a small meat-eating dinosaur from the small, sharp teeth in the jaws. The arms that support the flight feathers to form the wings still had three separate, clawed fingers. And *Archaeopteryx* has a long bony tail. All in all, this amazing little animal looks a lot like *Compsognathus*, which was celebrated for more than a century as the smallest dinosaur. Because of its combination of features, *Archaeopteryx* was recognized as the link between reptiles and birds. And by 1870, Thomas Huxley publicly announced that birds had descended from dinosaurs. This became the prevailing belief for the end of the 19th and the early part of the 20th century.

A Danish bird specialist, Gerhard Heilmann, undertook a very thorough study of bird origins, publishing his conclusions in English in 1927.[2] He felt that dinosaurs were anatomically the closest animals to *Archaeopteryx* and other early birds. However, he pointed out that dinosaurs were probably not the direct ancestors of birds because they lacked certain features, or characters. For example, most of us are familiar with the wishbone in chickens and turkeys that we eat. A wishbone is formed by the fusion of a pair of bones found in most vertebrates, from fish to mammals (in humans we refer to them as collarbones). But dinosaurs, according to what Heilmann knew at that time, did not have a collarbone. And if dinosaurs had lost the collarbone, how could they be ancestors of a group of animals that still had a wishbone made of collarbones? Once a bone is lost through evolution, it is virtually impossible to get it back. Therefore Heilmann concluded that birds and dinosaurs were probably "cousins" with a common ancestor. This conclusion became the prevailing hypothesis of bird origins for the next half century.

What Heilmann did not know is that a small dinosaur with a wishbone had been discovered in Mongolia in 1923. Unfortunately, the wishbone, which was almost identical in shape to that of *Archaeopteryx*, had been

misidentified as a different bone. This mistake was not corrected until almost 50 years later by a Polish scientist. Today, we know that many meat-eating dinosaurs had wishbones, including the giant tyrannosaurs. In many cases the wishbones had been collected with the skeletons, but had been misidentified as ribs or other bones.

A resurgence of interest in the question of bird origins occurred in the 1970s, and three theories prevailed. Some scientists maintained that Heilmann was correct and that birds came from a relatively primitive group of reptiles that were generally known as "thecodonts." This group included the ancestors not only of birds, but also of crocodiles, flying reptiles, and dinosaurs. Other scientists resurrected the idea that birds are the direct descendants of dinosaurs. And others believed that birds came from primitive crocodiles. The last theory is not as unreasonable as it may sound initially, because birds are in fact more closely related to crocodiles than they are to any other living animals, and some early crocodiles were in fact small animals that may have run around on long hind legs and even climbed trees.

New dinosaur discoveries towards the end of the 20th century led to the establishment of dinosaurs as the most likely bird ancestors. This change was assisted by the introduction of a more precise method of classifying living animals and plants. Known as "cladistics" or "Phylogenetic Systematics," large numbers of derived or specialized characters are analyzed by computer and compared (see Holtz, page 31). Meat-eating (theropod) dinosaurs were found to share with birds more than 125 unique characters that are not found in any other extinct or living animals.[3] Although the majority of paleontologists came to accept that dinosaurs were the most likely ancestors of birds, some maintain that this was not possible, and most of the public is still unfamiliar with the evidence.

What if Dinosaurs Are Ancestral to Birds?

At the same time that we started to visualize birds as dinosaur descendants, a theory was be-

ing debated concerning whether or not dinosaurs were warm-blooded. This debate has not been resolved, but there is a preponderance of evidence to suggest that small theropods (meat-eating dinosaurs) did have a type of physiology similar to modern birds and mammals (see de Ricqlès, page 79). The combined ideas that birds may have come from warm-blooded dinosaurs allowed some perceptive paleontologists to predict that one day we would discover dinosaur fossils with feathers. The reasoning is simple. No matter which group of animals birds arose from, fully-formed flight feathers could not have sprouted from the arms to form wings—there must have been intermediate structures that could subsequently be adapted for flying. Feathers, therefore, must have been present in some form in the ancestors of birds.

What could feathers have been used for if they were not there for flight? That answer is easy because we know that feathers, especially down, have remarkable insulation properties. If, as some scientists assumed, birds evolved from small theropods and small theropods were warm-blooded animals, it is possible that small theropods had evolved feathers as a way of staying warm. By the early 1980s, some artists had extended the idea of dinosaurs feathered for warmth to dinosaurs using feathers as ornaments. Even though many of us who worked on dinosaurs felt that these ideas were logical (and probably correct), I do not think that most of us ever expected to see fossil proof within our lifetimes. It is difficult to find the fossils of small, feathered theropods because feathers are part of the soft anatomy, which usually rots away without being fossilized. Also, small dinosaurs are much less common than large dinosaurs, whose bones are large enough to resist destruction by scavengers and decomposition.

The Feathered Dinosaurs of Asia

Even though the number of skeletal characters shared by dinosaurs and birds is the strongest evidence of their close relationship, many have resisted the idea that something like a robin could evolve from anything related to *Tyranno-*

Fig. 2.
Sinosauropteryx is a small theropod with an extremely long tail and short arms. The fossils of this animal are well enough preserved to include simple, feather-like structures (see arrows).

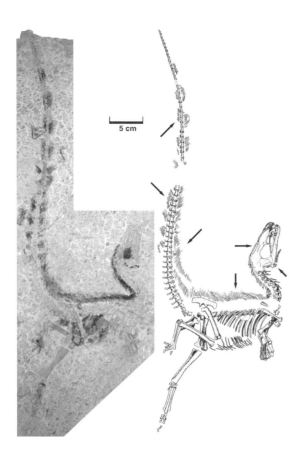

5 cm

saurus rex. When one prominent paleontologist was asked in 1996 what would convince him that the theory was correct, he stated that a dinosaur would have to be found with fossilized feathers. Several months later, I was in China when a Chinese newspaper reported just such a discovery. With some of my Chinese colleagues, I went to see the specimen with the expectation of seeing fossilized fungus, mineral crystals, or something else that had been misinterpreted as feathers. But it only took seconds to be convinced that the beautifully preserved, 125 million year old dinosaur really did show evidence of some kind of downy or hairy covering. Approximately the size of a large chicken, *Sinosauropteryx* (Fig. 2) seems to have been covered from the top of the head to the end of the tail by slender filaments. The filaments are stacked on top of each other, so it is difficult to see exactly what a single structure looks like. However, the evidence favors them being a simple branching structure similar to a down feather with a thick stalk at the base, and long slender filaments towards the outside (Fig. 3).

Three additional specimens of *Sinosauropteryx*, all showing the same kind of body covering, have now been recovered from the same locality in northeastern China, along with more than 1,000 fossilized birds.

The discovery of *Sinosauropteryx* focused the attention of scientists and public alike on the question of bird origins, and triggered a great deal of controversy. Controversy in science is not a bad thing of course, because it stimulates additional research. Before long, another "feathered" dinosaur was found in northeastern China. This one was named *Protarchaeopteryx*. Whereas *Sinosauropteryx* had short arms and a very long tail, *Protarchaeopteryx* had almost the opposite proportions. Its body was also covered with some simple, downy covering. However, at the end of the tail there are long stiff feathers that cannot be distinguished from the tail feathers of a mod-

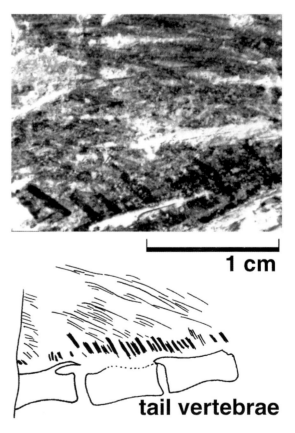

1 cm

tail vertebrae

Fig. 3. Photograph and drawing of the feathers along the tail of *Sinosauropteryx*, showing the short central stalks below, and the long filamentous structures above that branch off them.

ern bird. The feathers have a stiff central shaft (Fig. 4), and the vane is formed of the Velcro-like barb and barbule arrangement characteristic of bird feathers. More specimens of this animal were reportedly found later in 1997, and I went back to China to help my colleagues, only to discover that the fossils represented yet another new species.[4] *Caudipteryx* (Fig. 5) is similar to *Sinosauropteryx* and *Protarchaeopteryx* in having a downy body covering, and like the latter, it has long stiff feathers at the end of its short bony tail. In addition, this small dinosaur has long, stiff feathers behind its arms. Microscopically, these feathers are indistinguishable from those of modern birds.

The Chinese locality where the feathered dinosaurs are found has turned out to be one of the most remarkable and productive dinosaur sites in the world. The conditions more than 125 million years ago in northeastern China were perfect for the preservation of feathers. The bodies of the birds and feathered dinosaurs were washed into shallow lakes, where they would settle to the bottom. There they were gently buried by mud and volcanic ash under chemical conditions that seemed to prevent bacteria from decomposing the feathers.

At least three more species of Chinese theropod dinosaurs have been found with feathers in recent years. *Beipiaosaurus* is the largest feathered form found so far, and is as large as a human. It is part of a relatively strange group of theropods known as therizinosaurs. These unusual animals had relatively small skulls with leaf-shaped teeth that indicate dietary preferences for fish or possibly even plants. Their bodies were bulky, but the hands and feet were armed with strong, sharply tapering claws. Related to *Velociraptor* and *Deinonychus*, two other Chinese theropods, *Sinornithosaurus* and *Microraptor*, have also been found with preserved feathers. The Gobi Desert of Mongolia has now also produced theropod fossils with evidence of feathers. *Nomingia* is related to *Oviraptor*, but has a relatively short tail that ends in a pygostyle (sometimes called the "pope's nose" in a roast chicken or turkey), which is a fused series of five vertebrae that

supports the tail feathers and preening gland in modern birds. *Shuvuuia* is related to *Mononykus*, and is considered by some to be a bird, and by others to be a dinosaur. Remnants of feathers were preserved with one of the specimens.

The number of specimens of feathered dinosaurs recovered since 1996 is truly amazing. Although all are theropods, they represent

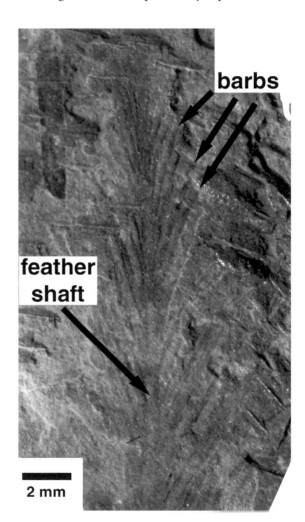

Fig. 4. Tail feather of *Protarchaeopteryx* showing the central shaft and barbs. The barbs are more or less parallel because of the smaller, Velcro-like structures called barbules that hold them together.

many different families that are as different from each other as the modern carnivore families of cats, dogs, bears, weasels, civets and hyenas. The diversity strongly suggests that feathers were present in most of the Upper Cretaceous dinosaurs of the Northern Hemisphere. For example, even though feathers have never been seen on any specimen of *Oviraptor* from the Gobi Desert, nor on *Deinonychus* from Montana, the fossils of close relatives of these animals were well enough preserved to

have feathers. *Caudipteryx* is closely related to *Oviraptor*, and *Sinornithosaurus* to *Deinonychus*, which suggest that feathers may have covered these animals too. In fact, looking at the distribution of feathers amongst theropod dinosaurs even opens up the possibility that a giant animal like *Tyrannosaurus rex* may have had feathers on its body at some stage of its life! Feathers are unlikely to have covered its body at maturity, however, as six-ton terrestrial animals do not need insulation.

"feathers."[6] In this animal, the body appears to have been covered by simple-branching, down-like feathers. They were not very long, but would have been perfect for insulating the body. Covering the body with insulation strongly supports the idea that small theropods were warm-blooded, or at least had a much more active metabolism than lizards and crocodiles do.

Small endothermic animals generally have more problems than large ones in keeping their

Fig. 5.
Caudipteryx is a small dinosaur from northeastern China that has long bird-like feathers behind its arms and at the end of the tail, and simpler, downy feathers over much of the rest of its body.

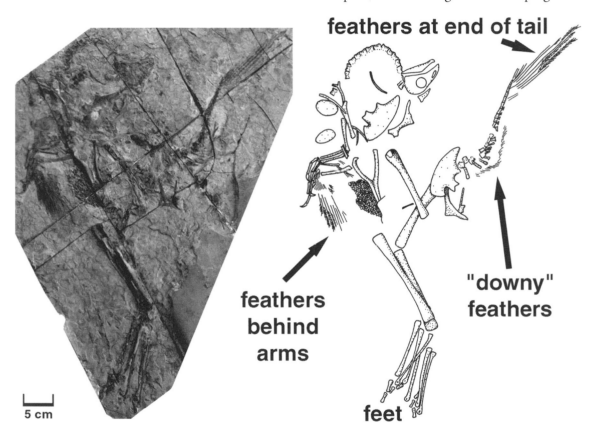

The Evolution of Feathers

Scientists have been speculating about the origin of feathers for almost a century and a half. Some agree that they were probably derived from reptilian scales, but even this idea is under debate. The driving evolutionary mechanism has been even more elusive. The discovery of feathered dinosaurs has provided a framework within which feather evolution must fit,[5] although there is still considerable latitude for scientific speculation and debate. *Sinosauropteryx* is the most primitive theropod dinosaur that we know of with evidence of

body temperatures constant. This relates to size scaling differences. To use a simple example, let's assume that *Sinosauropteryx* could fold up into a cube that was one meter by one meter by one meter. Its body surface area, through which heat is lost, is proportional to one meter squared. The volume and mass of the same animal would scale with one meter cubed, and this would be proportional to the amount of heat produced within the body. In this animal the amount of heat lost through the skin (surface area) compared to the amount of heat produced would be proportional to $1^2/1^3$, which is

the same as 1/1. An animal double the length would have a heat loss to heat production ratio of $2^2/2^3$, which is 4/8 (= 1/2). In other words, doubling the size of the animal reduces heat loss to about half the rate. Taking this ratio to an extreme, let's assume that *Tyrannosaurus rex* is 30 times the length of *Sinosauropteryx*. The heat loss ratio of this giant would be $30^2/30^3$, equivalent to 900/27,000 (= 1/300). As a very rough estimate then, *Tyrannosaurus rex* lost body heat at a rate of about 1/300th that of *Sinosauropteryx*. To use a living example, heat loss is the reason a shrew has to eat its own body weight in food every day, while an elephant eats only a fraction of its body weight.

Fluctuations in body temperature are generally not a problem in large animals, but can be problematic in small ones like *Sinosauropteryx*. And this is probably why feathers have only been found on small dinosaurs. But insulation works in two ways; it can keep an animal warm, or if that animal is cold, the insulation can prevent it from warming up. It would make no sense for a cold-blooded animal to be insulated, because it relies on external sources like the sun to warm up. An ice cube insulated by down takes much longer to melt on a hot day than an ice cube exposed directly to the sun. Similarly, a cold-blooded insulated animal would need to stay warm, because once it did cool down, it would take much more effort to warm up again.

Protarchaeopteryx and *Caudipteryx* show us the next stages in the evolution of feathers. Long, stiff, complex feathers that are indistinguishable from the contour and flight feathers of modern birds appeared at the end of the tail and behind the forearms. Initially, these may have developed as a way of displaying to potential mates and rivals. Dinosaurs in general were very visual animals that developed a tremendous array of crests, frills, horns, plates, spikes and scales for display. Most of these features are only preserved in the skeletons. Once feathers were present on dinosaurs, however, they seem to have been quick in utilizing them for display. Although such a display hypothesis is only one possible interpretation, it does make good sense. Feathers are lightweight, strong, colorful, and replaceable, making them better

than most of the display structures developed from bone. Spreading out a fan of tail feathers or lifting the arms was probably as effective for theropods to attract mates as it is in modern birds.

Nests of *Oviraptor* and related forms have been found in Mongolia and China[7]. In a half dozen cases, adults have been found fossilized on top of the nests. The eggs were laid in a circle, two at a time, as the mother stood on one spot and turned her body. At the same time as she rotated and laid eggs, she used her hands to scoop sand onto the eggs. We can infer this because in one of the specimens, the hands are stretched out beside the nest and are at a lower level than the eggs. However, the feet in the space are level with the top of the eggs. The result was a mound nest with the eggs laid in a spiral as many as three layers deep. The mother dinosaur remained on the nest, with her feet in the center where there were no eggs. Her chest covered some of the eggs, and her tail covered others. But in all cases, the arms are wrapped around the outside of the nest. The eggs between the body and the outspread arms appear to have been unprotected. However, if the *Oviraptor* had long feathers projecting behind the arms like its close relative *Caudipteryx*, then the feathers would have protected those eggs[8]. This idea has led some scientists to speculate that the elongate feathers behind the forearms and hands may have evolved, perhaps in conjunction with the display function, as a way of protecting the eggs in a nest.

Once long, stiff complex feathers were present behind the arms and at the end of the tail, they probably gave these small theropods an aerodynamic advantage over their rivals. Perhaps they flapped their arms as they jumped across logs or ditches, and that simple action gave them just enough speed or control to escape a predator or capture a prey item. Selection would then favor lengthening those feathers in subsequent generations, and would also control the development of new ways to utilize the arms. Ultimately the descendents would have become airborne for at least short distances, and a whole new world would have opened up to these dinosaurs. It was not a

simple or easy transition because it involved the development of complex rearrangements of bones, joints, muscles and many other things. But many of the changes that led to the sophisticated flight apparatus of modern birds are documented in the fossil record.

The Origin of Flight

There has been a lot of speculation over whether the active flight of birds developed in tree-climbing dinosaurs, or ground-running species. Many animals have taken to the air, including insects, fish, amphibians, reptiles and mammals. Some of these animals are active fliers that are able to stay in the air for prolonged periods of time by actively generating thrust. This category includes most insects, flying reptiles (pterosaurs), bats and, of course, birds. But there is also a diverse assemblage of gliding animals, which includes flying frogs that use the membranes between their toes for parachuting, lizards (and even snakes!) that extend their ribs to form gliding membranes, and flying squirrels that have membranes stretched between their arms and legs. Some paleontologists believe that birds started flight by jumping out of trees, just as most of these gliders do[9]. Others argue that the complex arm movements of a bird must have started in a ground-running dinosaur, and that flight began from the ground up.[10] Good arguments support both ideas, although the long-legged, small, feathered theropods of China were mostly ground-dwellers and therefore support the "ground up" theory. Without knowing the precise point at which active flight began, it is perhaps impossible to know the answer. Perhaps the correct answer was even a combination of both the "ground up" and "trees down" theories. Even though we may never know which theory is correct, the lively debate has motivated scientists to study and better understand the mechanisms involved in the evolution of active flight.

Are Dinosaurs Extinct?

In spite of the large amount of public debate that has gone on in recent years concerning the origin of birds, the overwhelming evidence shows us that birds evolved directly from theropod dinosaurs. No other candidates exist with so many anatomical similarities and new evolutionary features (including feathers) as are shared between birds and dinosaurs. Under a modern biological and paleontological classification, birds are a subset of the Dinosauria. This means that they can be referred to as living dinosaurs, and that dinosaurs are not extinct. In a sense, dinosaurs are still more successful than mammals because there are more than twice as many living species. Birds are very specialized dinosaurs, which we will continue to call birds. But knowing that this one lineage of dinosaurs did survive the great extinction event of 65 million years ago does give dinosaurs a new edge of respectability. As the dominant land animals for more than 140 million years, dinosaurs were long considered to be one of the most magnificent failures in the history of the planet. We can still marvel about why so many dinosaur lineages disappeared at the end of the Cretaceous, but that must now be tempered with the idea that dinosaurs are still successful animals.

Somehow knowing that birds are living representatives of the Dinosauria also gives birds a new respectability. I have never been able to look at a robin in quite the same way as I did before knowing that its bloodlines merged with those of *Tyrannosaurus rex* and *Velociraptor*!

Summary

Feathers are not unique to birds. Fossilized feathers have been found in at least seven species of non-avian theropod dinosaurs (*Beipiaosaurus, Caudipteryx, Microraptor, Protarchaeopteryx, Shuvuuia, Sinosauropteryx, Sinornithosaurus*). Because these animals represent a great diversity of different families (Fig. 6), we can infer that many, many other species of dinosaurs had feathers when they were alive.

The first function of feathers was not for flight, but for insulation in small, warm-blooded, non-avian dinosaurs. It is possible that initially the insulation was restricted to the small, temperature vulnerable young. Because evidence suggests that all dinosaurs hatched from eggs, none of which were more than half a meter (18 inches) long, no newborn dinosaur

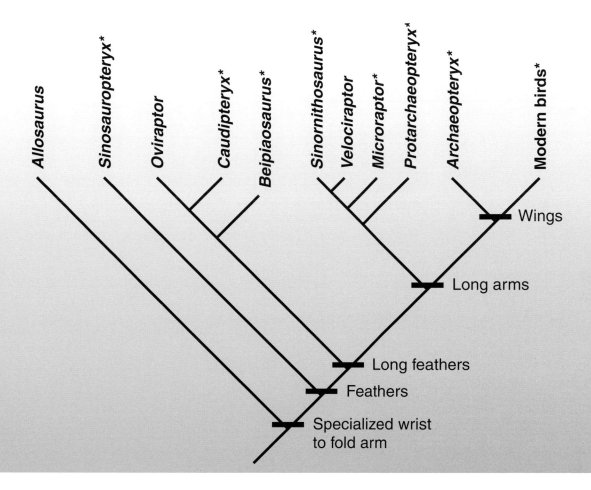

Fig. 6.

Cladogram showing the interrelationships of the feathered dinosaurs and modern birds, along with the stages at which certain avian characters were acquired. *Allosaurus* is included as an example of a famous dinosaur that we do not believe had feathers.

babies would have been longer than a meter (3 ft). Insulation would not have been necessary in the adults of large species, but may have been retained in the adults of small species. Once feathers were present in dinosaurs, they were adapted into many other functions. Long stiff feathers at the end of the tail and behind the arms may have been used initially for displaying to potential mates, and/or for protection of eggs in the nest. The presence of these long feathers gave some of these dinosaurs aerodynamic properties that could be used in prey capture and escape from enemies. This selective advantage would have been emphasized by lengthening the feathers and arms until active flight was possible. Flight requires many other modifications in anatomy, physiology and behavior, all of which would have been changing at the same time as the wings developed.

Active flight is what separates birds from their non-avian theropod ancestors. However, because there is a continuous series of evolutionary steps between birds and their ancestors, it is difficult to know exactly when active flight began. Paleontologists therefore often arbitrarily consider *Archaeopteryx* as the division between non-avian theropods and birds, because this animal has been proclaimed the "first bird" since 1861. Using this scheme, any animal that shares a common ancestor with *Archaeopteryx* and modern birds would be considered as a bird. Theropods more primitive than this common ancestor would not be considered as birds, even if they had feathers and rudimentary wings.

Under a modern biological and paleontological classification, birds are a subset of the Dinosauria. Thus, dinosaurs are not extinct, and are still one of the most successful groups of vertebrates. In spite of the fact that birds are dinosaurs, they are also a very specialized group that we can continue to call birds.

References

1. Hecht, M.K., J.H. Ostrom, G. Viohl, and P. Wellnhofer, eds. 1985. *The beginnings of birds*. Proceedings of the International *Archaeopteryx* Conference, Eichstatt, 1984. Freunde des Jura-Museums Eichstatt, Eichstatt, Germany.

2. Heilmann, G. 1972. *The origin of birds*. New York: Dover Books (reprint of D. Appleton & Company edition of 1927).

3. Gauthier, J. 1986. Saurischian monophyly and the origin of birds. In *The origin of birds and the evolution of flight*, ed. K. Padian, 1–55. San Francisco: California Academy of Sciences.

4. Currie, P.J. 1998. *Caudipteryx* Revealed. *National Geographic* 194(1):86–89.

5. Prum, R.O. 1999. Development and evolutionary origin of feathers. *Journal of Experimental Zoology* 285:291–306.

6. Currie, P.J. 1997. Feathered dinosaurs. In *The encyclopedia of dinosaurs*, eds. P.J. Currie and K. Padian, 241. San Diego: Academic Press.

7. Dong Z.M., and P.J. Currie. 1996. On the discovery of an oviraptorid skeleton on a nest of eggs at Bayan Mandahu, Inner Mongolia, People's Republic of China. *Canadian Journal of Earth Sciences* 33:631–636.

8. Hopp, T., and M. Orsen. 1998. Dinosaur brooding behavior and the origin of flight feathers. In *Dinofest international symposium, program and abstracts 27*, eds. D.L. Wolberg, K. Gittis, S. Miller, L. Carey, and A. Raynor. Philadelphia: Academy of Natural Sciences.

9. Chatterjee, S. 1997. *The rise of birds*. Baltimore: Johns Hopkins University Press.

10. Padian, K. and L.M. Chiappe. 2000. The origin of birds and their flight. In *The Scientific American book of dinosaurs*, ed. G.S. Paul, 190–202. New York: St. Martin's Press.

Dinosaur Extinction: Changing Views

J. David Archibald
Department of Biology
San Diego State University, San Diego, CA

12

I f you played a word association game with people and asked them to respond with the first word that popped into their heads, more than likely, the response for "fossil" would be "dinosaur." If you asked them "what killed the dinosaurs" more than likely they would say "an asteroid." Just as there are many unanswered questions about how dinosaurs lived, so too there are many unanswered questions about how they died. Before we can address the specific question of dinosaur extinction, we must examine the general topic of extinction.

Extinction: No One Gets Out of Here Alive

Before the beginning of the 19th century the idea that species could become extinct was not widely believed. In the western culture, it was believed that all species of plants and animals were perfectly created in a matter of days. Certainly a creator would not allow his creations to disappear from Earth. The work of the renowned French paleontologist, Georges Cuvier soon showed that some species have disappeared. He compared the teeth of the two species of living elephants with those of fossil elephants and their relatives. The teeth of fossil mastodons and mammoths were so different from those of the two living species that Cuvier concluded that the mastodons and mammoths no longer existed. Very soon the work of Cuvier and others showed that not only had extinction occurred, but that it had been a very common event throughout Earth's history.

Scientists even began to use extinction to measure geological time. Probably best known for this kind of work is the English geologist Charles Lyell. Lyell compared collections of modern marine invertebrates (clams, snails, etc.) to collections of fossil marine invertebrates from different aged rocks and places in Europe. He found that the older and more different the collections of fossil invertebrates were from modern invertebrates, the more extinction had occurred.

> *. . . it appears that marine regression, an asteroid impact, and massive volcanism each probably played a significant role in what is the best known mass extinction in Earth's history.*
>
> *Dave Archibald*

David Archibald is Professor of Biology and Curator of Mammals at San Diego State University. He received his Ph.D. from the University of California, Berkeley in 1977. He has written numerous articles, essays, reviews, and monographs on the systematics and evolution of early mammals, biostratigraphy, faunal analysis, and extinction. His fieldwork has taken him from the American West to Middle Asia. His 1996 book *Dinosaur Extinction and the End of an Era: What the Fossils Say* (Columbia University Press) documents what we know of the fossil record at the time of dinosaur extinction.

Lyell was measuring the fact that, over long periods of geological time, as species evolved others became extinct.

Today, we know that not only does extinction occur, but that of all the species that ever lived, well over 99 percent are now extinct. This figure is not an exact percentage, but rather it is an approximation, based on three factors: (a) the age of the Earth, (b) the number of species alive at any given time, and (c) an estimate of how long the average species exists.

a) Paleontologic evidence points to the origin of life on Earth at about 3.5 billion years.

b) For the number of species alive at any given time, we can use estimates of E.O. Wilson, who placed the number of known species alive today at 1.4 million, while judging that the number may be anywhere from 10 to 100 million.[1] To be conservative, we can use the lower estimate of 10 million for the number of species alive at any given time. This number overestimates the number of species alive early in the history of life, but underestimates the number of species alive later in geological time.

c) In the late 1980s Niles Eldredge estimated how long a variety of species survived. An average duration for a species based on his various estimates was 12.4 million years.[1]

We now need to make a few calculations. If we use the estimate that the number of species at any given time is about 10 million, that each species lasts about 12.4 million years, and that life originated 3.5 billion years ago, we arrive at the staggering figure of over 2.8 trillion species that have lived on Earth! If only 10 million are alive today, this means that only one in every 280,000 species that has ever lived is alive today. Thus, 99.99 percent are extinct. This total may seem like a shocking number, but it is clear that the fate of most species is extinction rather than further evolution. Because most species become extinct, only a small percentage of species provides the future diversity. Extinction is so common that it is the rule rather than the exception. Just as evolution is an ongoing process adding new species, extinction is an ongoing process that eliminates species.

Armed with this information, we might think that the extinction of species that we humans are causing today is normal or even common. Such an assumption would be wrong. The great many extinctions occurring today because of human activity fall into an extremely rare category known as mass extinction. In the past 540 million years, during the Phanerozoic Era, there have been five times when the numbers and rates of extinction became so high that they stand out from all other times in the geological past (Fig. 1). A sixth mass extinction is now underway, but this time humans are the culprits.

In Fig. 1, each point on the graph represents the number of extinctions per million years for families of marine invertebrates and vertebrates for that particular interval of geological time. The solid line is called a regression line and can be thought of as showing the average extinction rate through geological time. The dashed lines surround 95 percent of all the points shown on the graph. Each of the five spikes falls well outside the 95 percent interval, indicating much higher rates of extinction five times in the past. These five spikes are generally recognized as times of mass extinction.

Fig. 1.

The five major mass extinctions in the past 540 million years as demonstrated by extinctions of marine invertebrates and vertebrates. From Archibald (1996) after Raup and Sepkoski (1982).[1]

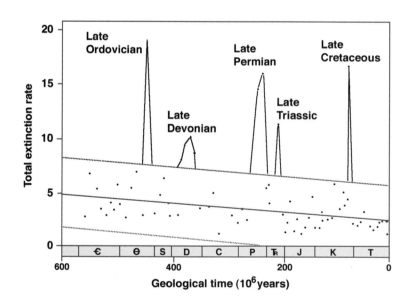

When we add up all the extinctions that have occurred in the past, we find that mass extinctions probably account for no more than about 10 percent of all extinctions. The other 90 percent or more are normal or background extinctions that are the counterpart to evolution. Although the five mass extinctions comprise a relatively small percentage of total extinctions during Earth history, each represents a major reorganization of the Earth's biota. The severest of the big five reorganizations occurred at the end of the Permian some 250 mya (million years ago). There was over 90 percent species extinction, although this is not obvious from Fig. 1.

The most famous mass extinction, however, is the one that included the last of the dinosaurs, the terminal Cretaceous mass extinction 65 mya.[2] This mass extinction wiped out the dominant land vertebrates and opened the evolutionary way for mammals that until that time were no larger than a small cat. Although the terminal Cretaceous mass extinction is the most famous, we still cannot say what happened with certainty in part because of the relatively poor record of its best-known victims—the dinosaurs.

The K/T Extinction

The Late Cretaceous (from about 100–65mya) is not only the last geological time interval from which dinosaurs are definitely known, it is also one of the best sampled for dinosaurs. The name K/T comes from a combination of the names Cretaceous, in which we find fossils of dinosaurs, and Tertiary, the time in which we see the first appearance of modern groups of mammals (the K comes from Kreide, the German word for "chalk," deposits of which are common at this time in Europe; Cretaceous means "chalk-bearing").

Around the world there are literally hundreds of Late Cretaceous dinosaur sites, including such seemingly unlikely places as Antarctica and New Zealand (Fig. 2, A). One of the best-known areas for Late Cretaceous dinosaurs is in the northern part of western North America. In this area, we know that the number of genera of dinosaurs dropped from

32 to 19 during the last 10 million years of the Late Cretaceous. In other words, some 40 percent of dinosaur genera were lost in this region during that time. Whatever killed the last species of dinosaur, the record shows that they were declining during the last 10 million years

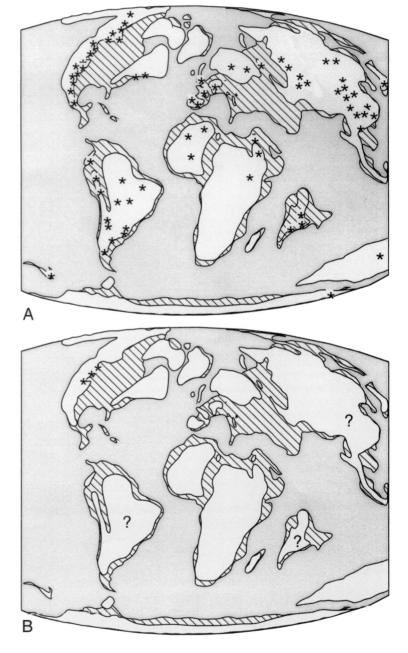

Fig. 2. A. Important dinosaur sites (black stars) for the Late Cretaceous (100 to 65 mya). **B.** Important vertebrate faunas spanning the K/T boundary. More recently discovered possible K/T boundary vertebrate faunas are indicated by a question mark. From Archibald (1996)[1] using data mostly from Weishampel (1990) and map after Smith et al. (1994).

of the Cretaceous, at least in the western part of North America.

Unfortunately, as one examines the rocks representing the end of the Cretaceous, the fossil record of dinosaurs gets worse. Only a handful of places in the world have a nearly continuous fossil record of vertebrates across the K/T boundary. The one area where we do have an adequate record of dinosaurs very near the K/T boundary is in western North America. Recently discovered sites in China, South America, and India offer hope in the future for an even better K/T record of vertebrates (Fig. 2, B).

What this means is that the vertebrate (and dinosaur) fossil record at the K/T boundary is far poorer than is usually realized. Most importantly, this means that at least for now, we cannot say anything about how fast dinosaur extinction occurred. The record is simply too poor to address this issue. Fortunately, the record is good enough that we can say something about how many species of dinosaurs and other vertebrates became extinct and how many survived. We can also say something about what is called the selectivity of these extinctions. This means that we can ask whether all kinds of species of vertebrates from bony fish to turtles were equally or unequally affected during the change from the Cretaceous to the Tertiary. In order to ask such questions we need a fossil record of vertebrate species other than dinosaurs. Indeed, we have such a record, once again in the western portion of North America. As we examine this record, keep in mind questions such as: Why did this dinosaur extinction occur? Was it only the dinosaurs that went extinct? Which other groups of animals survived?

Other Vertebrate Species Living with the Last Dinosaurs

Our K/T sites for vertebrate fossils are found in eastern Montana. It has been well documented that at least 107 species of vertebrates existed here during the closing million or so years of the Cretaceous. These species belong to 12 major groups, which are listed in Table 1. It may be surprising that of these 107 species

only 19 are dinosaurs. You may also notice that pterosaurs (flying relatives of dinosaurs) and birds are not included in this table, even though fossil evidence indicates their existence at this time.[3] These two groups are not included because their fragile skeletons make them poor candidates for fossilization compared to most other vertebrates. Of the 107 remaining species of vertebrates, 49 percent (52 species) survived, or 51 percent became extinct.[1] Even though only about half of these species of vertebrates became extinct, the overall biological effect on the land was profound. Dinosaurs had been the dominant land vertebrates throughout the Mesozoic. Although mammals had been around for almost the same amount of time, it was only after the dinosaurs were gone that mammals truly began to flourish. More importantly, we can use the survival data in Table 1 to test the various theories of dinosaur extinction.

The Three Most Popular Theories of Dinosaur Extinction

Soon after the discovery and naming of the first known dinosaurs in the early 19th century, people began to speculate on what had happened to them. By the 1980s, there were over 80 dinosaur extinction theories, more than for any other group of animals. With so many theories, it is no wonder that they ranged from the highly reasonable to the absurd, e.g., overhunting by aliens.[4, 3] Given so many theories of extinction, only the three best explained and testable of these will be discussed here. All three theories, to one degree or another, argue that changes in climate, whether sudden and drastic or slow and cumulative, caused the extinctions. It is important to keep in mind that none of the three theories is mutually exclusive, that is, any combination of the three may have happened. The most important question is which theory, or part of a theory, is best supported by the fossil evidence. Before testing these theories against the vertebrate fossil record, they first must be described. They are described in the order of the length of the time intervals over which they are thought to have acted, beginning with the longest.

The Deccan Traps

Massive eruptions of flood basalts on the Indian subcontinent, called the Deccan Traps, occurred at the time of the K-T boundary. Flood basalts flow from great fissures and volcanoes with moderate amounts of explosive power. They are more like the lava flows of the Hawaiian Islands than the explosive eruption of Mount St. Helens, which literally blew the mountain apart. However, the Deccan Traps erupted over four or more million years and produced enough lava to cover both Alaska and Texas to a depth of 2000 ft (610 m). The effects of such massive volcanism have not been as well studied as the effects proposed in other extinction theories. It is clear, however, that such eruptions would greatly increase the amount of very fine-grained material in the atmosphere. This "dust" would decrease the amount of sunlight reaching the Earth's surface, which would in turn lead to long-term global cooling. Both cooling and a decrease of sun reaching the Earth's surface would, over this long time interval, change the vegetation and thus, affect the animals feeding upon it.

Marine Regression

The second theory relates extinction to marine regression, which is the process whereby very shallow seas that covered much of the low-lying areas of continents drain away, back into the deeper ocean basins.[1] One of the greatest such marine regressions is recorded in rocks near the end of the Cretaceous Period, some 65 mya. Unlike the eruption of the Deccan Traps, which took place over millions of years, the terminal Cretaceous marine regression occurred over a much shorter period of time, only tens or hundreds of thousands of years. Estimates suggest that 11.2 million square miles of land were exposed during this interval, more than twice the next largest such addition of land during the past 250 million years. The landmass that was exposed is approximately equal to the area of modern day Africa. As these continent-sized shallow seas drained away, great areas of low, coastal habitat were fragmented into smaller and more isolated areas. As these habitats for terrestrial coastal spe-

Table 1. Survival and extinction of vertebrate species across the K/T boundary in the western part of North America.

Sharks and relatives	0/5	(0%)
Bony fish	9/15	(60%)
Amphibians	8/8	(100%)
Mammals		
Rodent-like multituberculates	5/10	(50%)
Placentals	6/6	(100%)
Marsupials	1/11	(9%)
Reptiles		
Turtles	15/17	(88%)
Lizards	3/10	(30%)
Crocodile-like champsosaurs	1/1	(100%)
Crocodilians	4/5	(80%)
Dinosaurs		
Ornithischia (bird-hipped)	0/10	(0%)
Saurischia (reptile-hipped) except birds	0/9	(0%)
Total number and percent survival	52/107	(49%)

The first number is the number of species that survive the K/T boundary extinction. The second number is the number of species known from the Late Cretaceous. Percent is percent survival.

cies shrank and became more distant from one another, population sizes would have decreased. Furthermore, as land emerged from the sea, land bridges were exposed, such as the Bering land bridge between Asia and North America. This new land would allow migration of terrestrial vertebrates and the potential for increased competition among previously separated species. River systems that had once flowed over relatively short distances grew in length as the shoreline receded further and further and provided greater habitat for many fresh water organisms. As new land areas were exposed with the regression of the great interior seaways, the climate cooled and climatic extremes increased, further stressing an already stressed environment.

Asteroid Impact

The argument presented in this theory is that a 6-mile wide asteroid struck Earth 65 mya, spewing very fine material high into the atmosphere where it spread around the globe.[5] The major result was the blockage of many of the sun's rays. This blockage greatly reduced or possibly stopped photosynthesis. Many indi-

vidual plants would have been stunted or killed, and many plant species would have become extinct. Herbivorous dinosaurs and other vertebrates that fed upon these plants would have disappeared, which in turn would have caused the extinction of the carnivorous dinosaurs that fed upon the herbivores. This process appears to have taken only a few thousand years at the most.

Evidence for the asteroid impact comes from three sources:

The crater The probable crater for this impact has been located near the tip of the Yucatan Peninsula in Central America. It was appropriately named Chicxulub, which means the devil's horns in a local dialect. At 110 miles (180 km) across, it was originally thought to be the second largest such structure on Earth. More recent studies argue that it is more on the order of 60 mi (100 km), or possibly even smaller.

An increase in the element iridium at the Cretaceous/Tertiary (K/T) boundary A high level of iridium found in rocks from this time period in many places on the Earth is a strong indication of an extraterrestrial source for the iridium, such as from an asteroid striking the Earth. Iridium, a very heavy element, like gold, is rare at the surface of the Earth except where concentrated in a small area by very deep volcanoes.

Shocked quartz Quartz grains, showing shocked lamellae (or parallel layers) in two directions at a microscopic level, indicate a great amount of rapidly applied pressure—such as might be caused by the impact of an asteroid.

Following the publication of the asteroid impact theory in 1980, a number of other possible consequences were suggested, such as acid rain, globe wildfire, sudden temperature increases and/or decreases, tsunamis, and superhurricanes. Each of these is argued to have had consequences on the world's plants and animals.[5] Some of these consequences can be tested using the fossil vertebrate record, but others cannot, notably tsunamis, and superhurricanes. We now turn to examining and testing the various theories using mostly information from the fossil record of vertebrates.

Using the Vertebrate Record To Test Theories of Dinosaur Extinction

The most obvious pattern of extinction among the major vertebrate groups listed in Table 1 is that extinctions were concentrated in only five of these groups: sharks and their relatives, lizards, marsupials, ornithischian dinosaurs, and saurischian dinosaurs. Species in these five groups account for 75 percent of the extinctions. This pattern demonstrates that the K/T extinctions were highly selective and any theory of extinction must account for this selectivity. Because the biotic effects of volcanic eruption have not been explored extensively, but the effects are considered to be similar to the effects of an asteroid impact, these two theories will be discussed together under the asteroid impact.[1] We will start with the marine regression theory.

Marine Regression

Global marine regression began in the last few million years of the Cretaceous. As this occurred, tremendous new tracts of dry land were added. Dinosaurs may well have lived away from the seacoast near the end of the Cretaceous, but their habitats would not have been affected. The well-known dinosaur-bearing vertebrate localities near the K/T boundary, however, come from coastal plain habitats. Thus, it is from this information that we should draw our inferences of what may have occurred.

As indicated in Table 1, the fossil record shows a 0 percent survival for both dinosaurs and sharks and their relatives. With marine regression, the coastal plain habitats were being drastically reduced, stranding dinosaurs in ever-smaller areas—this is similar to what humans are doing to the habitats of large mammals in Africa today. The loss of habitat stressed the dinosaur populations, setting them up for any other biotic insults such as that from even a small asteroid impact or from massive volcanism. At the same time, the coastlines were retreating away from the Western Interior taking the sharks and relatives with them.

Sharks could follow freshwater courses up to a few hundreds of miles or kilometers, but not thousands of miles or kilometers. Their marine connections were severed. Much larger and longer rivers replaced the small coastal streams, which continued to support many species of freshwater fish, turtles, amphibians, and crocodilians. This too is supported by the evidence shown in Table 1. In fact, freshwater species did very well, with descendants such as paddlefish, sturgeon, gar, snapping turtles, and alligators still plying the Missouri-Mississippi river systems.[1]

The lowering of sea level reconnected once separated landmasses, such as eastern Asia and western North America. The fossil record shows that the earliest relatives of what would later evolve into hoofed mammals and whales probably reached North America at this time (65 mya)—their possible ancestors being placental mammals known in Asia 20 million years earlier. These new North American placental mammals had teeth that resembled those of the opossum-like marsupials living in North America at this same time. The marsupials arose in North America over 100 mya and were very common for at least the 20 million years leading up to the K/T boundary 65 mya. It seems likely that the appearance of these new placental mammals in North America spelled competitive doom for the marsupials. Interestingly, when both groups appear in South America a few million years after the K/T boundary, they do not compete. Rather, the placentals became more strictly herbivorous, while the marsupials became omnivorous and carnivorous, including large saber-toothed marsupial cats. The one group whose fossil record cannot be explained by marine regression is the lizards, which underwent a drastic reduction, at least in western North America. A possible explanation is that when the climate became wetter in this area following the K/T boundary, the more dry-adapted lizards could not tolerate the changes.

Asteroid Impact

We can start testing the asteroid impact theory by examining the effects of such possibly re-lated events as acid rain, sharp temperature decrease, and global wildfire.[5]

a) We know from work on living species and habitats that among vertebrates, acid rain hurts aquatic organisms most, killing both adults and eggs laid in the water. Among the aquatic organisms, however, only sharks and their relatives show very high levels of extinction. Other aquatic species did very well through the K/T transition, thus acid rain was probably not a major factor.

b) If a sharp drop in temperature had occurred, the species that should have been most affected would have been cold-blooded (ectothermic) vertebrates that spend at least part of their time out of water—this is why today we see far fewer species of amphibians and reptiles (except warm-blooded or endothermic birds) in the far northern and far southern regions of the world. Yet, most of these ectotherms, except lizards, did well through the K/T boundary. Whether dinosaurs should be considered as endotherms, ectotherms, or as having another kind of physiology remains controversial (see de Ricqlès, page 79).

c) Finally, a global wildfire is argued to have consumed 25 percent of all above ground burnable material. Geological evidence for wildfire has been presented based on large amounts of carbon and other compounds occurring at the K/T boundary. Some paleontologists argue that such a global wildfire would have transported great quantities of detritus (very small fragments of plants and animals) into the streams, which would have favored the survival of aquatic animals that eat such material.[4] Other scientists argue that a global wildfire would have been an equal opportunity killer: terrestrial creatures would have been burned on land, and aquatic vertebrates would have suffocated from all the burned material dumped into their environments. Thus, depending upon how the fossil information is interpreted, global wildfire could have been either a significant or unimportant event at the K/T boundary.

Volcanism and Asteroid Impact

One probable result of an asteroid impact or high levels of volcanic activity would have been

the blocking of sunlight, either around the globe or in more restricted areas. Blocking sunlight would have caused the reduction or even cessation of photosynthesis among green plants. The fossil record of land plants in the northern part of western North America suggests an extinction of at least 80 percent, which tends to support the hypothesis that photosynthesis was suppressed. Reduced photosynthesis would have had a devastating effect on large herbivores, especially if they were already stressed by other events such as marine regression. Unfortunately, more recent studies have questioned just how much dust really would have been spread around the world from such an impact. Thus, more studies are necessary before we can resolve this question.

Summary

When using the vertebrate fossil record to test the various theories of dinosaur extinction, it appears that marine regression explains more of the highly selective pattern of extinctions and survivals through the K/T transition in western North America than do either an asteroid impact or massive volcanism. When combined with evidence from plants and marine species, it appears that marine regression, an asteroid impact, and massive volcanism each probably played a significant role in what is the best known mass extinction in Earth's history. These three theories, plus other causes that we still do not know, each may have been necessary, but clearly were not enough individually to cause the extinctions that we see at the end of the Cretaceous.

References

1. Archibald, J.D. 1996. *Dinosaur extinction and the end of an era: what the fossils say.* New York: Columbia University Press.

2. MacLeod, N., and G. Keller (eds.). 1996. *The Cretaceous-Tertiary mass extinction: biotic and environmental changes.* New York: W.W. Norton and Co.

3. Dingus, L., and T. Rowe. 1998. *The mistaken extinction: dinosaur evolution and the origin of birds.* New York: W.H. Freeman and Co.

4. Fastovsky, D.E., and D.B. Weishampel. 1996. *The evolution and extinction of the dinosaurs.* Cambridge: Cambridge University Press.

5. Alvarez, W.L. 1997. *T. rex and the crater of doom.* Princeton, NJ: Princeton University Press.

Fossil Forensics: From the Field to the Lab

IV

Death, destruction, and scientific discovery right in your own backyard. **Fiorillo** gives readers an introduction to taphonomy, the study of how organisms begin their journey to potential fossilization. He explains how bones can be altered after an animal dies and shows how close examination of dinosaur bones and how they are preserved can give us clues to the life and death of those animals.

13

The article by **Wright and Breithaupt** describes two types of trace fossils left by dinosaurs: coprolites (fossilized excrement) and footprints. They explain how dinosaur tracks and trackways are preserved, and how they are classified. Readers will learn how these trace fossils can tell us a great deal about the anatomy and behavior of the animals that made them, and, at the same time, gain an understanding of the limitations of using tracks to hypothesize about dinosaurs.

14

So you want to collect dinosaurs? Sounds simple: find a dinosaur fossil, dig it up, and away you go—fame and the *New York Times* science reporter await you! **Chure**'s article takes you through the process that paleontologists follow, from deciding where to search for fossils, to the day they arrive back in the lab. Prepare for hard rock, hard work, and, oh yes, did you fill out all the paperwork?

15

16

Remember the *Velociraptor* skeleton so stunningly revealed by the ground-penetrating radar in *Jurassic Park*? While we have not yet matched Hollywood's perfect technology, paleontology has certainly come a long way since the days when scientists were limited to a rock hammer and a chisel. The article by **Chapman** and colleagues looks at some of the modern tools, such as GPS (Geographic Positioning Systems) and CT (Computerized Tomography) scans, that help paleontologists locate fossils in the field and work with them in the lab to produce increasingly accurate and detailed reconstructions of dinosaurs.

Taphonomy: The Story of Death and Destruction

13

Anthony R. Fiorillo
Dallas Museum of Natural History
Dallas, TX

Some time ago, Harlo, our black Labrador Retriever, wandered into the neighbor's pasture and found a cow bone, which of course made his day. Tail wagging, he brought it back into our yard where I spotted him, happily gnawing on the bone and leaving tooth marks all over its surface. Being interested in taphonomy, the study of how organisms become fossils, I have oftentimes been intrigued when I find a fossil bone with tooth marks on it. I wonder about the behavior of the animal that did the chewing. Did the "chewer" kill the "chewee," or did the "chewer" just happen on the bone? Given my interest in such matters, I realized that I didn't have a picture of bone chewing in action. So I went back into the house, grabbed my camera and approached Harlo for a great taphonomy-in-action picture.

The point of this little story is not so much to publicize my fascination with my dog, as it is to use my dog as an example of how simple observations can lead to scientific thought. This story also illustrates important components in the subdiscipline of paleontology called taphonomy. Within vertebrate taphonomy, this study focuses on what happens after an animal dies and before it is buried, and how a particular bone might provide clues to reconstruct aspects of an ancient ecosystem.

> *Understanding the processes that can act on bones before burial is profoundly important in the discipline of taphonomy.*
>
> Anthony Fiorillo

In this case, Harlo's bone chewing illustrates several components of vertebrate taphonomy. First, definite changes can occur between the time an animal dies and the time that it is buried. A variety of biological and physical processes can act on bones before final burial. Here, at least one bone of the cow was being altered from its original form by Harlo's teeth.

When reconstructing past environments, we need to ask how representative a collection of fossil bones is of the actual ancient ecosystem? For example, Harlo's removal of the bone from the original carcass illustrates an important issue, transportation. Some bones are removed from an ecosystem, while others remain in place and are incorporated into the fossil record in their original site. Understanding the processes that can act on bones before burial is profoundly important in the discipline of taphonomy. These same processes can be observed while watching the family dog.

Anthony R. Fiorillo received his Ph.D. from the University of Pennsylvania and is currently the Curator of Earth Sciences at the Dallas Museum of Natural History. His research focuses on the taphonomy and paleoecology of dinosaurs and the paleoenvironments of dinosaur-bearing rock units. As a result of his research interests, Dr. Fiorillo has led numerous expeditions in western North America, particularly Texas, Montana, Wyoming, Colorado and Alaska, in addition to field work in Asia, Australia, and South America.

Historical Perspective

One of the early observers in vertebrate taphonomy was William Buckland, a geologist from the first half of the 19th century and truly one of the giants in the field of geology. Buckland was a gifted and talented geologist. To many, Buckland's fame stems from his 1824 description of Megalosaurus,[1] one of the first two dinosaurs introduced to the world. Of even more significance to those of us interested in bone-chewing behavior in animals, is Buckland's work on the Kirkdale cave fauna in 1823.[2] The Kirkdale cave is a rather unremarkable cave in a limestone unit in Yorkshire, England. Buckland found that it contained an assemblage of very young fossil bones—the remains of a variety of animals, such as tigers, hyenas, mastodons, and rhinoceroses. You might have noticed that none of the animals just listed live in England today; and their modern relatives live in warmer climates, far to the south.

What does any of this discussion have to do with Harlo and bone-chewing animals? Buckland recognized that some ancient predator had chewed many of the bones in the Kirkdale cave. By using modern animals as a model—in this case, hyenas from the zoo—he was able to establish that hyenas were the residents of the Kirkdale cave. Consequently, the bones of hyenas and their prey had not been transported into the cave from some great distance; rather, all of the animals actually were residents of the area. Based on this interpretation, Buckland was able to suggest that the climate of England had changed, or that these animals had adapted to an environment different from the one in which we currently find them. Thus, Buckland was the earliest observer of taphonomic processes, or at least the earliest to record such observations in print. As a dog owner, I wonder if Buckland had a dog and if so, was it the dog that tipped him off on the significance of the Kirkdale cave?

Taphonomy was originally defined as the study of information loss during an organism's transition from the biosphere to the lithosphere, or in other words the transition an organism makes from being a living entity to becoming a fossil. The term was originally de-fined by the Russian paleontologist I.A. Efremov in 1940,[3] but use of the term lan-guished until the 1960s. It was through his understanding of Russian, and his friendship with Efremov, that E.C. Olson, a paleontologist at the University of Chicago was able to bring the term to the literature written in English.[4] As a result, many American vertebrate paleontologists were introduced to taphonomy through the efforts of E.C. Olson. Starting in the mid-1970s, Anna K. Behrensmeyer launched the discipline of vertebrate taphonomy onto the path leading to modern studies.[5,6] Significant work has continued in taphonomy through to the present, but it is important to remember that the coining of the term did not invent the science, because paleontologists have recognized taphonomic patterns since Buckland.

Clues from the Bones

Bone modification can be loosely defined as features on bones that are the result of any process that occurs after death but before fossilization (e.g., trampling, scavenging, weathering) and that alters the shape of a once-living bone.[7,8] The resulting change to the bone is referred to as a bone modification feature. By this definition then, any pathological feature resulting from diseases such as arthritis, a bone altering disease that can be preserved in the fossil record, are excluded. Damage from crystallization is also excluded. Bones have a porous center that provides ample space for the formation of inorganic crystals during the fossilization process. Sometimes these crystals can grow beyond the available space and essentially "explode" the bone.

Bone modification processes can be divided into two groups. One group includes those processes that are active in a river environment, such as abrasion. The second group includes processes active on the land surface, such as weathering and trampling.

Bone weathering (Fig. 1) has received considerable attention, given its implications for taphonomic interpretations. Behrensmeyer defined bone weathering as "the process by which the original microscopic organic and inorganic components of bones are separated from each

other and destroyed by physical and chemical agents operating on the bone *in situ*, either on the surface or within the soil zone."[6] She defined six stages of the weathering of bone, based on factors such as bone cracking and flaking. These weathering stages have since been modified for direct application to fossil material.[9] Simple observational studies on fossils have shown us decided patterns to bone weathering. In general, the longer a bone has been exposed on the surface, the more well-developed the cracking and flaking on the bone. Consequently, we would expect that the more advanced the weathering stage in an assemblage of fossils, the longer it has taken to bury the bones and accumulate the fossil deposit.[10]

How then does this knowledge of accumulation time impact interpretations of ancient ecosystems? Some have argued that fossil deposits accumulate over sufficient lengths of time to record evolutionary change; others regard these accumulations as a snapshot of a paleoecosystem at a single instant in time. This issue was examined at Carnegie Quarry at Dinosaur National Monument, the most famous quarry in the world for Jurassic-aged dinosaurs. Here it was determined that the period of time for the accumulation of the dinosaur bones was on the order of a few months to perhaps 19 years—an ecological time scale.[11] Therefore, this quarry is truly an ecological snapshot, and the dinosaurs found here all existed in one ecosystem.

Bone Chewing

Another example of bone modification is tooth marks. Harlo is just one example of an animal that chews bones. A variety of animals other than canids chew bones, and they do so for a variety of reasons. Rodents, for example, chew

bones to wear down their teeth and to obtain nutrients from the bones themselves. As it turns out, with respect to mammalian predators, how bones get chewed tells us something about the behavior of the scavenger or predator that did the chewing. In a very interesting study some years ago, Haynes observed that at least some mammalian predators (bears, canids,

large cats, and hyenas) chew on the carcasses of their prey in a predictable manner.[12] For example, bears and hyenas are more inclined to break bone, while wolves are more inclined to leave gnaw marks on bone. It follows then, that if one knows the chewing patterns, one can determine the predator's identity from the tooth marks on a particular carcass; an interesting tool when working with very recent assemblages of fossils. In at least one example from a 14 million-year-old Miocene locality, the specific identity of both a mammalian carnivore and its prey could be determined based on tooth marks (Fig. 2).[9]

With respect to terrestrial predators in the Mesozoic, it has been suggested that theropods were similarly capable of chewing.[13, 14] Tooth-marked bone from Mesozoic fossil assemblages, however, is decidedly less common than tooth-marked bone from Tertiary fossil assem-

Fig. 1.
A. Sheep jaw from eastern Montana showing little to no cracking or flaking from weathering.
B. Sheep jaw from eastern Montana showing extensive cracking and flaking from prolonged exposure to weathering.

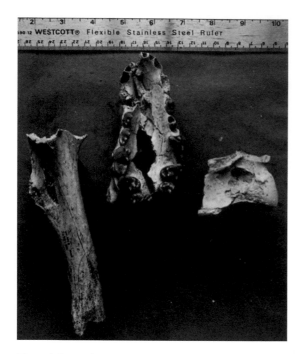

Fig. 2. The predator and its prey? A fossil canid (*Aelurodon*) skull and two horse limb bones from a Miocene fossil assemblage in southwestern Nebraska. Notice the puncture marks on the two horse bones. These puncture marks match precisely with individual teeth from this canid, indicating the identity of the animal that made the marks. See Fiorillo (1988) for details of the taphonomy of this site.

blages.[8,15] This difference suggests that predatory dinosaurs did not routinely chew the bones of their prey. Further, the few reports of tooth-marked bone in dinosaur assemblages generally are single, shallow tooth grooves on bone surfaces, rather than extensive gnawing. These observations would indicate that the marks were the result of accidental contact between theropod teeth and the bones of prey, rather than active gnawing as in mammalian predators.

Bone Abrasion

A third bone modification feature commonly recorded in taphonomic studies is the degree of bone abrasion. Simply put, a bone fresh from a carcass is a certain size and shape. If that particular bone is placed in a river and transported downstream, gradually the movement of the bone against the stream bottom will abrade the bone. If this abrasion continues for some

lengthy interval of time, the bone can be reduced to an unrecognizable pebble. Some taphonomists have formalized the degrees of abrasion with specific stages, based on recognizable features of wear.[16, 9, 17] Identification of the amount of abrasion on bones is helpful in trying to distinguish between bones that were buried near where an animal lived, versus bones that might have been transported from some place far removed from the final burial locality. This issue can be important when trying to determine which fossil animals might have lived together in the same ecosystem.

Clues from the Assemblage

In addition to the wealth of attention paid to bone modification features, features of the bone assemblage itself are also useful clues in reconstructing how a particular fossil deposit formed. What follows are standard features typically discussed in analyzing the taphonomy of any particular bone bed.

Bone Distribution

The three-dimensional distribution of fossils within a single rock unit is typically heterogeneous in terrestrial vertebrate fossil localities. However, concentrations of fossils commonly occur in the bottom portion of a rock unit. Such a distribution is typical of a mass mortality of animals, all being buried quickly. In other words, these assemblages represent very short periods of time, that of an "ecological snapshot." In contrast to these assemblages, bones distributed throughout a rock unit, rather than concentrated, typically suggest a slower rate of accumulation for the fossil assemblage. Thus, these assemblages of bones represent much larger segments of time and may not be indicative of a single ancient ecosystem.

Skeletal Articulation

Another important feature is the degree of articulation of a skeleton; that is, are the bones of an assemblage isolated skeletal elements or are whole skeletons preserved? Based on his study of dinosaur skeletons preserved in channel deposits at Dinosaur Provincial Park, Peter

Dodson argued that the degree of articulation of a skeleton is inversely related to the length of time between that organism's death and final burial;[18] that is, the longer the time between death and burial, the less likely that the skeleton will remain articulated. Others have substantiated this claim with studies of modern vertebrate carcasses, which have also determined, at least in mammal skeletons, that the relative amount of easily decomposable tissue and the joint type also contribute to how readily skeletons fall apart.[19, 20, 21]

Bone Transport

In taphonomic studies, resolution of the role of the physical environment in the formation of a given fossil assemblage is important in ecological reconstructions, as one needs to determine the amount of mixing of animals from different ecosystems (such as from long distance river transport) that might have occurred. With respect to an assemblage of isolated fossil bones, one of the first tasks is to determine whether or not the bones have been transported into the site by fluvial (river) currents. Behrensmeyer and Korth devised a means for testing the compatibility between bones and the surrounding sediment.[5, 22] Both Behrensmeyer and Korth assumed that the rate at which bones settle reflects the transportation potential of the bones. They measured the settling velocities of a wide range of bones and determined a relationship between bone settling velocities and grain sizes. This relationship provides the means for determining correspondence between bones and sediments at a fossil site.

They found that they were able to determine if the stream currents responsible for depositing the sediments at a given fossil site were sufficient in strength to also transport the bone into the site. If the currents were too weak to transport the bones, then the fossil animals found likely lived near the locality at the time the sediment was being deposited. For example, fine-grained material such as clay and mud are deposited in relatively quiet water. Dinosaur bones, as large as they are, need relatively strong currents for them to be trans-

ported. Therefore, if dinosaur bones are found in very fine-grained material, transportation is unlikely. The bones probably belonged to animals that lived near the site of burial.

In his now classic study, Voorhies noted the tendency of long bones such as ribs, limb bones, etc., to align parallel to a water current if the bones were submerged.[23] Similarly, MacDonald and Jefferson noticed a tendency for waterlogged tree trunks to align preferentially in a current.[24] The degree of alignment of linear bones is dependent on stream velocity, particle size, and shape. If long bones show an alignment, it is reasonable to conclude that a stream current acted on the assemblage of bones. In some fossil assemblages, this orientation of bones can be the only clue to such a taphonomic event.

Clues from a Biased Assortment

The last feature to be discussed here is designed to determine the relative strength of bias within a fossil assemblage. The variation in the size, shape, and density of the skeletal elements in the vertebrate body is reflected in the manner in which these bones behave in a stream environment.[23, 25, 10, 26] The lighter, less dense elements tend to be carried away sooner than the heavier, more dense ones. Voorhies' original concept of tallying skeletal elements at a site can serve as an effective means of establishing whether or not that site had been subjected to a sorting mechanism, either by streams or by a biological process such as scavenging. Straightforward tallying of skeletal elements is sometimes not the optimal method, since Haynes showed that mammalian carnivores can selectively remove parts of the prey skeleton.[27] Behrensmeyer and Dechant Boaz provide a means for estimating the degree of fluvial sorting by calculating tooth/vertebrae (T/V ratios).[28] Vertebrae tend to be among the first skeletal elements removed by stream action, whereas teeth tend to be among the last.

Voorhies experimented with skeletons in a flume and noted a consistent pattern of hydraulic sorting. He generalized his experimental data to predict that, in nature, the elements of the mammalian skeleton might be dispersed by fluvial currents as three hydraulically compatible groups, generally summarized as: ribs and vertebrae, limb bones, skull and jaws. A census of skeletal elements present in a fossil assemblage compared to the predicted numbers of bones would be an indicator of the sorting to which the assemblage had been subjected.

The predicted number of bones is based on the number of individuals found at a site. In a simplified example, take an assemblage of bones that has skulls as the most common element. Suppose 10 skulls are present. If these particular animals had 200 bones in their individual skeletons, then the predicted number of bones for the site is 2000. Next let's assume that all available bones in the assemblage are excavated and collected. After the excavation is completed only 1000 bones are recovered. Therefore only half the expected number is present. Knowing something about exactly which bones are present provides insights into why only some bones are present. For example, if the 1000 bones recovered are all the heavy, denser bones, it seems likely that intermediate water currents would remove the lighter bones, such as the ribs and vertebrae. If instead the elements removed were limb bones, this would tell us a different story. Such limbs generally

have a good deal of flesh attached, and most likely were removed by scavengers. Obviously tooth-marked bones in the assemblage strengthen such a conclusion. Going back to my dog, Harlo, we have a great example of just such a scenario: he scavenged the bone, removing it from the rest of the carcass.

Time Averaging

Paleontology is unique as a science devoted to the study of life on Earth, as it incorporates a scale of time unlike any other scientific discipline. Much of the previous discussion focused on the methods for determining if an assemblage of bones represented the fauna of a single ancient ecosystem or multiple faunas from several ecosystems. Implied in this latter determination is that all of the ecosystems were operating at the same time. But what about the issue of the mixing of faunas from different times?

A river eroding its banks, exposing bones previously buried, which are then redeposited down river, is a mechanism for mixing faunas from different times. Perhaps a more everyday example of combining material from different time planes is to go back to the bone Harlo

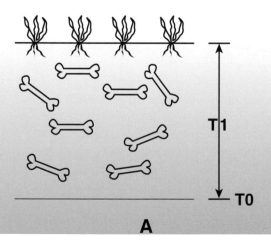

A

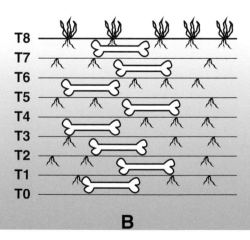

B

Fig. 3. Models for the sequence of events for the formation of a fossil bone accumulation. Model A shows a single depositional event (T1) after the bones achieved the appropriate weathering stage. This model provides the minimum amount of time needed for the formation of the quarry. Model B shows a cycling of depositional events (T1–T8) occurring. One cycle consists of a bone weathering, being buried, and vegetative cover developing. This second model shows the sequence of events needed for a maximum amount of time needed for the formation of a deposit. These models were applied to the Carnegie Quarry at Dinosaur National Monument for the estimates of time presented in the text (Fiorillo, 1994).

brought into the yard. This bone is from a cow that died a number of years before Harlo brought it into the yard. The death of the cow represents one time plane, while the introduction of the bone into the yard occurred during a later time plane. This combination of events from different time intervals is referred to as time-averaging.

The scale of the spacing between time planes determines the resolution in studying ecosystems through time. Therefore, an appreciation of the effects of time-averaging is important when trying to reconstruct an ancient ecosystem. Estimates for time-averaging are usually derived from reasonable estimates of factors such as rates of bone weathering, soil formation, and river dynamics.

The estimation of time for accumulation of dinosaur bones at the Carnegie Quarry, Dinosaur National Monument, ranged from a few months to 19 years (Fig. 3).[11] That theoretical minimum age of a few months was derived from bone weathering that, based on modern observations, took only a few months before bones achieved that particular level of weathering. The theoretical maximum age was derived from recognizing the maximum length of time that this particular weathering state could exist before entering the next stage—in this specific case, four years. The thickest bone in the assemblage is the one-meter tall pelvis of a sauropod dinosaur. The thickness of the bone bed is four meters. If the bones are superimposed as four very separate depositional events, i.e., floods, using bone weathering alone suggests an interval of 16 years.

But what of the time represented between depositional events? Given that this is an an-cient river channel setting, most rivers have one major period of flooding per year. Therefore, three planes, each representing a potential additional year, are to be added to the total. Thus, a theoretical maximum amount of time is 19 years. But again, the actual number of years is not the issue so much as the magnitude. This exercise shows clearly a logical estimate for invoking ecological time scales for the quarry at Dinosaur National Monument.

Summary

Taphonomy is the study of the processes that affect an organism between the time of death and fossilization. Though the term was formally defined in 1940, because the science is based on observational skills and experimental data, paleontologists have been making taphonomic observations for generations prior to the formalization of the discipline.

The intention here has been to highlight the power of observation. In virtually every component discussed it was simple observations in modern environments that formed the key to scientific advancement. Even from Buckland's earliest taphonomic experiment with hyenas at the zoo, observations of the patterns of bones found in modern ecosystems have provided great insights into the study of ecosystems of long ago. Taphonomy, is a powerful, multidisciplinary tool that has shed valuable insights into paleobiology, paleobehavior, paleocommunity structure, and the nature of fossiliferous depositional environments. Its potential is boundless, and it is based on simple observation and data gathering—perhaps even as simple as watching a dog.

References

1. Buckland, W. 1824. Notice on the *Megalosaurus*, or great fossil lizard of Stonesfield. *Transactions of the Geological Society of London* (2) 1:390–396.

2. Buckland, W. 1823. *Reliquiae diluvianae, or observations on the organic remains contained in caves, fissures, and diluvial gravel, and on other geological phenomena, attesting to the action of an universal deluge.* London: Murray.

3. Efremov, I.A. 1940. Taphonomy: new branch of paleontology. *Pan-American Geologist* 74:81–93.

4. Olson, E. 1962. Late Permian terrestrial vertebrates, U.S.A. and U.S.S.R. *Transactions of the American Philosophical Society* 52(2):1–224.

5. Behrensmeyer, A.K. 1975. Taphonomy and paleoecology of Plio-Pleistocene vertebrate assemblages east of Lake Rudolf, Kenya. *Bulletin of Comparative Zoology* 146:473–578.

6. Behrensmeyer, A.K. 1978. Taphonomic and ecologic information from bone weathering. *Paleobiology* 4:150–162.

7. Fiorillo, A.R. 1991a. Taphonomy and depositional setting of Careless Creek Quarry (Judith River Formation), Wheatland County, Montana, U.S.A. *Palaeogeography, Palaeoclimatology, Palaeoecology* 81:281–311.

8. Fiorillo, A.R. 1991b. Prey bone utilization by predatory dinosaurs. *Palaeogeography, Paleoclimatology, Palaeoecology* 88:157–166.

9. Fiorillo, A.R. 1988. Taphonomy of Hazard Homestead Quarry (Ogallala Group), Hitchcock County, Nebraska. *Contributions to Geology, University of Wyoming* 26:57–97.

10. Lyman, R.L. 1994. *Vertebrate taphonomy*. Cambridge: Cambridge University Press.

11. Fiorillo, A.R. 1994. Time resolution at Carnegie Quarry (Morrison Formation: Dinosaur National Monument, Utah): implications for dinosaur paleoecology. *Contributions to Geology, University of Wyoming* 30:149–156.

12. Haynes, G. 1980. Evidence of carnivore gnawing on Pleistocene and Recent mammalian bones. *Paleobiology* 6:341-351.

13. Farlow, J.O., D.L. Brinkman, W.L. Abler, and P.J. Currie. 1991. Size, shape and serration density of theropod dinosaur lateral teeth. *Modern Geology* 16:161–197.

14. Erickson, G.M., S.D. Van Kirk, J. Su, M.E. Levenston, W.E. Caler, and D.R. Carter. 1996. Bite-force estimation for *Tyrannosaurus rex* from tooth-marked bones. *Nature* 382:706–708.

15. Jacobsen, A.R. 1998. Feeding behaviour of carnivorous dinosaurs as determined by tooth marks on dinosaur bones. *Historical Biology* 13:17–26.

16. Hunt, R.M., Jr. 1978. Depositional setting of a Miocene mammal assemblage, Sioux County, Nebraska (U.S.A.). *Palaeogeography, Palaeoclimatology, Palaeoecology* 24:1–52.

17. Shipman, P. 1981. *Life history of a fossil: an introduction to taphonomy and paleoecology*. Cambridge: Harvard University Press.

18. Dodson, P. 1971. Sedimentology and taphonomy of the Oldman Formation (Campanian), Dinosaur Provincial Park, Alberta (Canada). *Palaeogeography, Palaeoclimatology, Palaeoecology* 10:21–74.

19. Toots, H. 1965. Sequence of disarticulation in mammalian skeletons. *Contributions to Geology, University of Wyoming* 4:37–39.

20. Hill, A.P. 1979. Disarticulation and scattering of mammal skeletons. *Paleobiology* 5:261–274.

21. Hill, A.P., and A.K. Behrensmeyer. 1984. Disarticulation patterns of some modern East African mammals. *Paleobiology* 10:366–376.

22. Korth, W.W. 1979. Taphonomy of microvertebrate fossil assemblages. *Annals of the Carnegie Museum* 48:235–285.

23. Voorhies, M.R. 1969. Taphonomy and population dynamics of the early Pliocene vertebrate fauna, Knox County, Nebraska. *Contributions to Geology, University of Wyoming*, Special Paper 1:69.

24. MacDonald, D.I.M., and T.H. Jefferson. 1985. Orientation studies of waterlogged wood: a paleocurrent indicator? *Journal of Sedimentary Petrology* 55:235–239.

25. Dodson, P. 1973. The significance of small bones in paleoecological interpretation. *Contributions to Geology, University of Wyoming* 12:15–19.

26. Blob, R. 1997. Relative hydrodynamic dispersal potentials of soft-shelled turtle elements: implications for interpreting skeletal sorting in assemblages of non-mammalian terrestrial vertebrates. *Palaios* 12:151–164.

27. Haynes, G. 1981. Prey bones and predators: potential ecologic information from analysis of bone sites. *Ossa* 7:75–97.

28. Behrensmeyer, A.K., and D.E. Dechant Boaz. 1980. The Recent bones of Amboseli National Park, Kenya, in relation to East African paleoecology. In *Fossils in the making*, eds. A.K. Behrensmeyer and A.P. Hill, 72–92. Chicago: University of Chicago Press.

Walking in Their Footsteps and What They Left Us: Dinosaur Tracks and Traces

14

Joanna L. Wright, *University of Colorado, Denver*

Brent H. Breithaupt, *University of Wyoming Geological Museum, Laramie*

D inosaur bones are body fossils—they were once part of the animal. Trace fossils, also called ichnofossils, are mostly sedimentary structures produced by animal behavior. They include, but are not limited to, footprints, burrows, and nests. Body fossils and trace fossils tell us different things. Body fossils tell something about how the animal looked in life and may enable us to estimate body size and mass. Trace fossils show what the animal actually did—for example, how it moved. They also can give us details about external features—shape of the pads on the feet, and even skin texture. By combining data from both sources, we can reconstruct a more complete picture of the lifestyles of prehistoric animals, such as dinosaurs, and work out how they did what they did.

Because trace fossils are produced by animal activity, they can also tell us things about the "tracemaker" that bones never could. Tracks, for instance, are preserved in place, so we know where the animal was when it was performing the activity. With bones, even when you find a complete skeleton, it could have been washed downstream for hundreds of miles. Aside from footprints, other dinosaur trace fossils that have been found include gastroliths and coprolites. Gastroliths are stones found in the abdominal cavity of some dinosaurs. They probably helped them to digest plant food by acting like grinding teeth in the stomach. Some living birds, such as chickens and turkeys, swallow grit to use in a muscular bag called a crop, which performs a similar function. But perhaps the most unusual types of trace fossil are the coprolites.

> *. . . tracks and trackways complement information from body fossils and help us to construct a more complete picture of life in the Mesozoic.*
>
> *Wright & Breithaupt*

Coprolites

These trace fossils are ancient feces, which preserve information on diet and defecation patterns. Coprolites often contain direct evidence of diet, and can

Joanna L. Wright is Assistant Professor in Geology at the University of Colorado at Denver. She received a B.S. (with honors) in geology from Imperial College, London and a Ph.D. in geology from the University of Bristol. Her research interests center around fossil terrestrial trackways, in particular locomotion and function, preservation, and paleoecology.

Brent H. Breithaupt is Director of the University of Wyoming Geological Museum, where he has worked for the past 22 years. He attended the University of Wisconsin-Milwaukee and the University of Wyoming. His research focuses on the history of vertebrate paleontology in Wyoming and the West and on the documentation and understanding of Mesozoic Era vertebrate faunas (including both trace and body fossils). He is a strong proponent of educational programs, public awareness, and partnerships in the field of paleontology.

provide important information about feeding levels and paleoenvironments. The biggest problem with this type of fossil is linking a particular animal to a specific coprolite.[1]

Researchers today use fecal remains to understand an animal's diet and dietary changes. Modern researchers have the capability to identify the animal that produced a particular fecal deposit. In the fossil record of dinosaurs, this task is very difficult to do, because we have no direct evidence of the diet of each species of dinosaur. Coprolites have been identified from various Mesozoic rock formations and have been attributed generally to herbivores or carnivores based on shape and content.[2] Although they are probably less common in the natural environment, carnivore feces are generally better preserved than those of herbivores, because they have higher calcium phosphate content, which results in a greater degree of mineralization.[1] Unfortunately, the morphology of coprolites can be highly variable, so shape rarely helps us identify the exact feces maker. However, we have been able to attribute some coprolites to specific animals, e.g., *Tyrannosaurus* and *Maiasaura*.

Dinosaur Tracks and Anatomy

The trace fossils that we are most concerned with in this chapter are dinosaur tracks. The word track can be used to mean either individual footprints, or a sequence of footprints. For this reason, we suggest that "tracks" not be used unless the meaning is absolutely clear. In this chapter, we will use the word "track" or "footprint" to mean a single impression of an animal's foot, whether a forefoot or a hindfoot, and we will use "trackway" to mean a sequence of such footprints made by a single animal at a certain time and place.

The study of tracks and trackways is more complex than it might first appear. We have to take into consideration several aspects of dinosaur anatomy and behavior: how they walked, how they placed their feet, how they stood, and the number of digits on each foot.

How They Walked

Dinosaurs may be bipedal, walking on their hind limbs only, like humans and birds; qua-

drupedal, walking on all fours, like cats or horses; or they can walk on two legs or all fours as their fancy takes them, like kangaroos or chimpanzees. This attribute is referred to as being facultatively bipedal.

How They Placed Their Feet

All tetrapod feet can be divided into two parts: the digits, toes or fingers, composed of bones called phalanges, and the rest of the foot, made up of bones called metacarpals or metatarsals. Dinosaurs walk on their toes, like cats and dogs, rather than flat-footed, like humans and bears. Thus, when we talk about a dinosaur footprint we are not talking about the impression left by the whole foot, but by parts of their toes and soft tissue structures such as fleshy "heel" pads. Sometimes dinosaurs left metatarsal (foot bone) impressions, usually when resting or walking under unusual circumstances.

How They Stood

Dinosaurs, like mammals and birds, have an erect stance. They hold their legs like vertical pillars under their bodies. When they move, their legs swing straight backwards and forwards. The opposite of this is a sprawling stance, such as in a lizard. Animals with a sprawling stance have to rotate their hips and shoulders.

Numbers of Digits

Dinosaurs may have three to five digits (fingers or toes) in each foot. As dinosaur trackers, we are only concerned with the number of digits that make contact with the ground in normal locomotion—their functional digits. Most theropods (meat-eating dinosaurs) have four digits on each hind foot, but all theropods are functionally tridactyl; that is, usually only three of their digits are recorded in their fossilized footprints. Many dinosaurs have a functionally tridactyl hindfoot. These groups include theropods, ornithopods and stegosaurs. However, most quadrupedal dinosaurs have a pentadactyl (five digits) forefoot, although it is often difficult to discern the separate digits in front footprints, especially in hadrosaurs and sauropods (Fig. 1).

As shown in Fig. 1, different types of dinosaurs leave differently shaped footprints and have different trackway configurations (the order and position of footprints). As a general rule, theropods are bipedal with a three-toed footprint, and ornithopods are facultatively bipedal with a three-toed footprint. Sauropods, on the other hand, are quadrupedal with five toes front and back. These toes are generally hard to see in the trackway so sauropod tracks are identified more easily by their great size, the characteristic oval hind footprint, and the horseshoe shaped front footprint. In contrast, ankylosaurs and ceratopsians have four-toed hind footprints and five-toed front footprints, but it is difficult to distinguish between the footprints of the two dinosaurs. The main difference is that ankylosaurs have a much longer outer toe (digit IV) than ceratopsians (Fig. 2).

Track Preservation and Classification

Tracks are made when an animal walks over the ground and exerts enough pressure to deform the surface. The amount of deformation is obviously related to the pressure per unit area that is transmitted through the animal's foot. This deformation, in turn, is related to weight, as large animals tend to exert more pressure per foot. However, other factors are also important, such as the relative size of the animal's foot, its weight distribution, and the moisture content and consistency of the sediment over which it walked. Tracks are best formed in wet and cohesive sediment, such as sand or mud. Tracks may be preserved as molds, the actual depression made by the animal's foot. They may also be preserved as casts, formed by the sediment that infilled the mold.

Most tracks are found in rocks that formed from sediments deposited beside the sea, on lakeshores, or near rivers. All of these areas would be intermittently wet and dry. The tracks formed when ani-

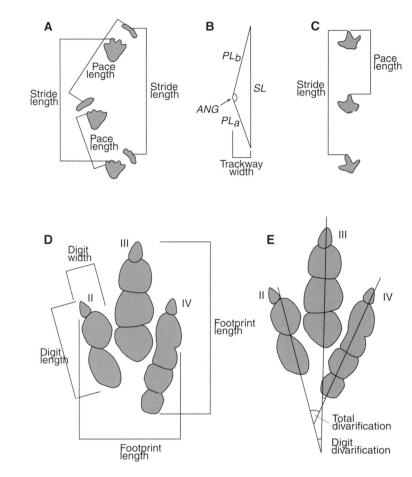

Fig. 1. Measurements of footprints and trackways. **A.** Pace length (PL) and stride length (SL) in a trackway of a bipedal dinosaur. Measurements can be taken from any point in the foot as long as the point used is specified and it can be easily identified on each footprint. For this reason the back of the heel or the tip of the middle toe are often used. **B.** The angles of the trackway angulation pattern; these are used as a measure of how in line the footprints are with one another. **C.** Pace and stride lengths in a quadrupedal trackway. Manus and pes trackways are treated separately. **D.** Measurements of footprint and digit lengths and widths. **E.** Measurement of divarification angles between the digits in a typical theropod track.

Fingers and toes are traditionally numbered from the inside to the outside and bones in the digits from the hand to the extremities. Thus, if you were to number the digits of your own hand the thumb would be digit I (roman numerals are always used for digits) and your little finger would be digit V. You have two bones in your thumb and three in all your other fingers. These bones are called phalanges (singular phalanx). Thus, the end bone in your thumb would be digit I, phalanx 2. A phalangeal formula is a shorthand for describing the number of bones and digits in a hand or foot and so the phalangeal formula for your hand would be 2,3,3,3,3 (assuming you have had no traumatic injuries or birth defects). Many dinosaurs have lost their inner and outer toe impressions. The pads in Fig. 1 show the phalangeal formula of theropods: 3, 4, 5.

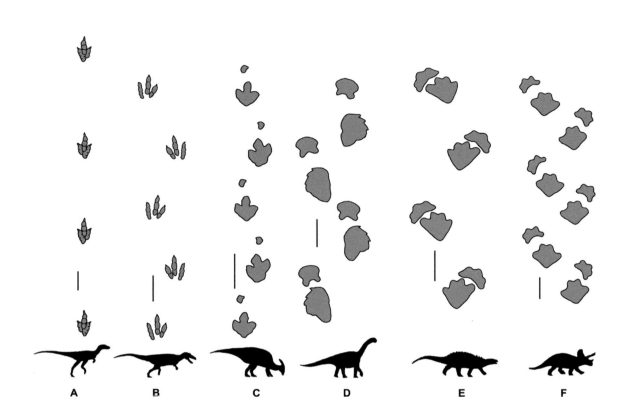

Fig. 2. Some common dinosaur trackways and their producers. **A.** *Grallator* trackway probably made by a small theropod such as *Coelophysis*. **B.** Large theropod trackway probably made by an allosaurid or megalosaurid. **C.** *Caririchnium*, large quadrupedal ornithopod trackway probably made by an iguanodontid or hadrosaurid. **D.** *Brontopodus* trackway probably made by a brachiosaurid/camarasaurid sauropod. **E.** Large quadrupedal trackway probably made by a nodosaurid ankylosaur. **F.** *Ceratopsipes* trackway probably made by a horned dinosaur. (B and E redrawn after Wright, 1996; C, D and F redrawn after Lockley, 1989)

mals walk over damp sediment. The sediment then bakes dry in the sun. Some clay-rich sediments become very hard when this happens, actually forming an adobe-like surface. When the tide next comes in, or the wind blows fine sand or silt, or when the river floods, the footprints are covered with more sediment and thus preserved. Preservation is even more likely to occur in damp environments where algae often form mats that bind the sediment together so that it resists erosion.

On the whole, track fossilization is a rare event. Many more tracks are made than are ever preserved. Sometimes the tracks might not bake hard in the sun; or perhaps they did bake, but wind and rain eroded them away before they were covered. Tracks might also be trampled by animals to the point that they are unrecognizable.

Some studies have shown that there is a zone around a lake where tracks are most likely to be preserved.[3] This model of the preservation of tracks could be called the Goldilocks Preservation Model: too near the lake, and the ground is so trampled that no discrete tracks can be discerned; too far away, the sediment is too dry and footprints don't get covered over by more sediment often enough. The zone between these two extremes, however, would be "just right."

When we find a dinosaur trackway that we think was made by a particular species of dinosaur, we do not give it the same name as that dinosaur. We give it its own name that reflects an aspect of the track morphology (shape). We do this because we usually cannot be absolutely sure which species of dinosaur made the tracks. Tracks are named using a classification system parallel to the classification of animals. Tracks are ichnofossils (trace fossils), and as such are given ichnogenus and ichnospecies names, and are sometimes placed in ichnofamilies. Just as

dinosaur names may often be recognized by the suffix "-saurus," trace fossil names can be recognized by suffixes such as "-ichnus," "-podus," or similar endings. The following common examples provide both the track names and their derivations:

Theropods
Eubrontes (true thunder),
Grallator (Grallae, i.e., heron/stork, -like)

Sauropods
Brontopodus (thunder foot)

Ornithopods
Caririchnium (track from Carir, Brazil)

Ceratopsians
Ceratopsipes (horned face [dinosaur] foot)

Ankylosaurs
Tetrapodosaurus (four-footed reptile)

It is not a good idea to name tracks to mean "tracks of a particular dinosaur," because further research may reveal that the tracks were made by a different animal, thus causing the name to be misleading. Tracks cannot be renamed once the original name has been published. Tracks are classified on the basis of shape. The age and size of the tracks do not matter. For instance, two tracks that cannot be distinguished by shape will be given the same ichnogenus or ichnospecies name, even if one track is from the Cretaceous of China, and the other is a track from the Triassic of North America. It is highly unlikely that the same animal survived for so many millions of years. Thus, trace and body fossils do not correspond exactly.

Estimating Speeds from Trackways

In the past 20 or so years, tracks have been studied much more intensively than previously.

We have made enough measurements on skeletons that we can now get quite a bit of information about the trackmaker from the size of its foot. The basic measurement of a dinosaur footprint is its length. The length of a dinosaur's foot (represented as FL) is related to the length of its leg up to the hip (represented as h). This ratio (FL/h) is different for different groups of dinosaurs, but as a general estimate, the hip height of a dinosaur can be estimated to be four times the footprint length. More accurate ratios for particular groups of dinosaurs have been worked out, as shown in Table 1.

Once you have a value for the hip height, and assuming you have a trackway with at least three footprints of the hind feet, you can start to estimate the speed of the trackmaker. The terms used in this explanation are illustrated in Fig. 1. Alexander made a series of observations on modern animals relating the length of their stride to the speed at which they were moving.[4,5] The relationship between stride length and speed was constant for animals as diverse as horses, ostriches, hedgehogs, and people, so it is likely that a similar relationship would hold true for dinosaurs. The speed can be worked out as relative speed, which is stride length (SL) divided by hip height (h). This allows you to compare the relative speeds of two similar animals. We also have an equation to convert pace, stride length, and hip height into a numerical speed in meters per second. This equation may give slightly low speeds for running animals, so if the trackway was made by a running animal, the second equation works best (Table 2).

How do you know if a trackway was made by a walking or running dinosaur? The difference between walking and running is that running has a suspended phase where all the limbs

Table 1. Estimates of hip height (h) from footprint length (FL) (after Thulborn 1990)[28]

Small theropods	*FL* < 25 cm	*h* = 4.5 FL
Large theropods	*FL* > 25 cm	*h* = 4.9 FL
Small ornithopods	*FL* < 25 cm	*h* = 4.8 FL
Large ornithopods	*FL* > 25 cm	*h* = 5.9 FL
Small bipedal dinosaurs in general	*FL* < 25 cm	*h* = 4.6 FL
Large bipedal dinosaurs in general	*FL* > 25 cm	*h* = 5.7 FL

Table 2. Relative and absolute speed estimates (after Alexander, 1989 and Thulborn, 1990)[5, 28]

Walk	$SL/h < 2.0$
Trot	$SL/h = 2.0\text{—}2.9$
Run	$SL/h > 2.9$
Speed of walking dinosaurs (m/s)	$V = 0.25g^{0.5}SL^{1.67}h^{-1.17}$
Speed of running dinosaurs (m/s)	$V = [gh(SL/1.8h)^{2.56}]^{0.5}$

Relative speed is based on the ratio between stride length (SL) and hip height (h). Estimates of actual speeds are extrapolated from the speeds and stride lengths of modern animals determined by observation. V is the speed of the trackmaker and g is the acceleration due to gravity. All linear measurements are in meters and all temporal measurements in seconds.

are off the ground at the same time. As a quick approximation, if the stride length is more than eight times the footprint length, then the dinosaur was running. To be more accurate, work out the relative stride length, and if that is greater than 2.9, the dinosaur was running. If the relative stride is less than 2, the dinosaur was walking. If the stride length falls between these two values, then the dinosaur may well have been trotting. However, trotting is an unusual gait for any animal, because it is very energetically inefficient. Interestingly, juveniles seem to trot far more often than adults do.

Trackways also allow some assessment of the way the animals were walking (Fig. 3). Were their limbs directly under their body? If so, their tracks should fall in a single straight line. Were their limbs under their shoulders? If so, there should two parallel lines of footprint in a single trackway. Trackways also show the position of the forelimbs relative to the hindlimbs—were they placed nearer or further away from the middle of the trackway? Such features can be characteristic of the trackways of certain animals. For instance, the trackways of small theropod dinosaurs often fall in a single straight line, whereas the trackways of large theropods and those of ornithopods are in a slight zigzag pattern with the toes often pointing inwards. Sauropod dinosaur track ways commonly have forefoot impressions falling inside the hindfoot impressions, whereas those of ankylosaurs and ceratopsians tend to have forefeet falling slightly outside the hindfeet (Fig. 2). Such features can help to de-

termine what kind of dinosaur made the trackway in question. Contrary to popular belief, trackways rarely show tail drags, and this is one reason why we now think that dinosaurs hardly ever dragged their tails.

Superlatives— Biggest, Smallest, Fastest

What were the biggest dinosaur tracks ever found? Sauropods, without a doubt, made the biggest tracks. Tracks from the Jurassic of Gansu Province, China, are reportedly the biggest ever found, with a maximum dimension of 1.5 m. These have not yet been officially described, so it is not clear if this dimension includes the outer footprint rim, in which case the foot might only have been about 1m long. Tracks from the Purbeck limestone group of the United Kingdom have been recorded with maximum dimension of 1.3 m, although these have only been described in detail in a National Trust Report, not in a scientific journal. This dimension also includes the external rim, so the dinosaur's foot was probably nearer to 1m long. Tracks from the Glen Rose site in Texas have been photographed with a toddler sitting in them, and maximum dimensions for them have been reported as 147 cm.[6] Again, this is a maximum footprint dimension and the actual foot length in this case has been calculated as 1.1m. Thus, most of the large sauropod tracks known so far indicate a foot length of slightly more than 1m. In all of these cases, the trackmaker was probably 20–30 m (66–98 ft) long and weighed in excess of 20–30 tons.

The smallest dinosaur tracks ever found are considerably smaller than the feet of any known adult dinosaurs. Even the diminutive *Compsognathus* would have left tracks 40 mm long, but we now know of several tracks only 25–30 mm long. These have been found in Jurassic rocks from North America,[7] and were probably made by juvenile dinosaurs.

Several tracks made by running dinosaurs are also known. The fastest estimated speed calculated from such trackways is 40 kmph.[8] Recently some tracks have been described from Jurassic age rocks in the United Kingdom.[9] These tracks were made by a large meat-eating dinosaur, approximately 6 m in length, probably weighing 1–2 tons and moving at 30 kmph (20 mph). This speed is much faster than people previously thought possible for these dinosaurs, and shows that the idea of these animals lumbering around all the time is very far from the truth. Similarly, biomechanical analysis of *Tyrannosaurus rex* indicates that this dinosaur could not have moved much faster than 25–40 kmph (15–25 mph).[10] *Tyrannosaurus* was much bigger and had longer legs, than the Jurassic trackmaker above, and larger animals tend to move relatively more slowly.

Theropods consistently seem to have moved faster than the herbivorous dinosaurs, and bipedal dinosaurs were probably speedier than quadrupedal ones. Many plant-eating dinosaurs were very large indeed, and some were so large that may not even have had to fear the meat-eaters. The fastest herbivores were probably the ornithopods, but no trackway of a large running ornithopod has been found, and very few tracks of smaller running ornithopods are known. One exception is the Lark Quarry site in Australia.[11] Trackways of running quadrupedal dinosaurs have not yet been documented.

Limitations of Information

As useful and versatile as tracks and trackways are, there are still some things that cannot be calculated from them. For instance, weight cannot be determined directly from the depth of the tracks. This observation may seem counterintuitive; after all, the depth of a given track is surely related to the weight of the ani-

mal that produced it, or at least by the pressure exerted by each of its feet. These factors are related, and in fact we can sometimes tell which dinosaurs are "front heavy" and which "back heavy" by looking at the relative depths of their front and back foot impressions. However, the

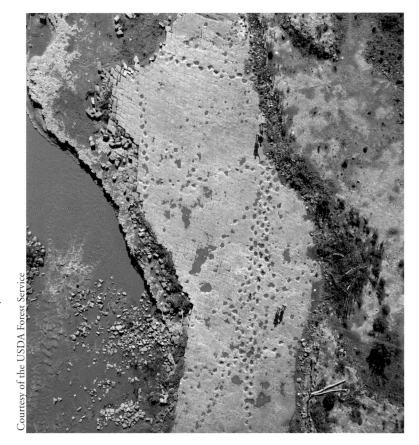

Courtesy of the USDA Forest Service

Fig. 3. Photo of sauropod trackways from a blimp-mounted camera at the Picketwire Canyonlands Dinosaur Tracksite in the Purgatorie Valley in southeastern Colorado.

depth of the tracks is also related to the consistency of the ground, so a heavy animal might leave shallow footprints on a hard substrate, while a much lighter animal would leave deeper footprints in a softer substrate.

Beaches illustrate this complexity well. Beaches have several zones of sediment consistency, which continuously change with the tides. The zone furthest from the sea is made of sand that remains dry, except when it rains and perhaps during storms. If you walk in this sand, your footprints will be a few centimeters deep. Nearer the sea, the sand may have a crust, which you break through; here you may leave similar footprints, but they will have a more jagged edge. Lighter animals would not break through such a crust, and would probably not leave any tracks at all. You may then encounter a zone of underpressured sand. This looks smooth and hard, but when you step on it, your foot sinks in as far as your ankle. Underpressured sand is common in the intertidal zone. The sea has retreated, but the sand has not yet settled. It is still precariously balanced in the same configuration as it was when the water was supporting the grains. As you step on it, the sand cannot support your weight and compacts to a more stable arrangement. Even nearer to the sea, the grains are still supported by water, and you will leave very shallow, but clear footprints. A simple experiment like this shows you the range of detail that may be seen in a footprint and also indicates the variety of factors we need to consider when interpreting fossil footprints.

Another common problem we face in interpreting trackways is the question of whether small tracks were made by juveniles or by adults of small species. Usually this question is not possible to answer. However, in two scenarios a juvenile trackmaker is the more likely interpretation. The first, and more convincing, is when the tracks are so small that no known adult dinosaur was small enough to have made them. The second instance is when both small and large tracks of the same morphology are preserved on the same surface, especially if the tracks seem to show gregarious behavior. It has been argued that this association of footprints would more likely represent juveniles and adults of the same species, than a coincidental assemblage of two or more species of adult dinosaurs in association.[12]

As mentioned earlier, most dinosaur trackways show normal walking behavior, but sometimes, we find a trackway that shows different kinds of behavior.

Limpers

Several trackways made by limping dinosaurs are known from different localities around the world. We can tell that the dinosaur was limping if one pace length is considerably and consistently shorter than the other.[13]

Squatters

Some dinosaur trace fossils were made by resting animals. They seem to have squatted down on all fours, leaving impressions of their hands, their feet and ankles, their noses, their pubic bones, and sometimes even the bottom of their tails. Impressions of this sort are known for at least two different species of small dinosaur.[14,15]

Swimmers

Some dinosaur tracks have been interpreted as those of swimming dinosaurs.[16, 17, 18] Some of these are probably either partial tracks or trace fossils made by other animals such as crocodiles,[19] but the jury is still out on some of these traces. It seems likely that dinosaurs could have swum, most animals can; we just do not have any solid evidence of this activity.

At some sites no distinct dinosaur tracks can be seen, but the entire surface is very uneven. This surface may indicate that the whole area was trampled by dinosaurs, resulting in the overlap of so many footprints that no individual ones can be distinguished. This phenomenon has been termed dinoturbation.[20] In fact, it has recently been suggested, by analogy with modern large animal game trails and migration paths, that the compaction caused by dinosaurs may have had a huge influence on the landscape, even to the extent of dictating the course of some river channels.[21]

Some myths and misconceptions surround tracks and trackways. One recurring myth is that there are places in the world where footprints of humans have been found with dinosaurs. Absolutely no evidence supports this. All of the sites where this is alleged to have occurred have been shown to be a result of misinterpretation of poorly preserved tracks, a deliberate hoax, or both. The most famous of these sites is the Paluxy River tracksite in Texas. The so-called "man tracks" were huge, did not show any impressions of distinct toes, and bore only a slight resemblance to the shape of a human foot. The tracks were followed further along the riverbed where it became clear that they were actually the tracks of a large meat-eating dinosaur. The "man tracks" had been

formed when the deep, narrow toe impressions of these elongate tracks had slumped in, leaving only the broad, shallower metatarsal (foot) impressions.[22] This interpretation was a case of mistaken identity, although some of the natural trackways found at this site had been embellished with artificially carved tracks in the shape of huge human feet.

Current and Future Research

Tracks have been used to increase our knowledge of the distribution of different groups of dinosaurs in space and time.[14, 23] Many rock formations do not contain bones, but do preserve numerous tracks. These tracks are the only evidence recording the presence of dinosaurs at these localities. Tracks have also enabled us to challenge some earlier paleoenvironmental interpretations. For example, the dinosaur tracks discovered in a unit of Wyoming's Sundance Formation indicate that this unit was not always submerged under a shallow sea, as had been previously thought.[12]

One of the more recent and encouraging developments in ichnology is the use of tracks to increase the accuracy of dinosaur reconstructions. Tracks have been used to demonstrate that ceratopsians could not have had a fully sprawling forelimb stance,[24] and to show that the hands of iguanodontids and hadrosaurs were directed slightly off to the side rather than straightforward as shown previously.[25] Tracks also indicate gait patterns of dinosaurs and have shown how gait changes with size or speed.[26, 9]

Computer animation programs are now available to test locomotion and trackmaking hypotheses, and have already been used with some success.[27] These programs can also be used to model unusual trackmaking situations—to try to explain strange tracks or trackways. Computers have also been used for some time to make accurate topographic maps of footprints. The hope is that these footprint maps will allow more accurate and objective determinations of the differences between individual tracks and trackways. The extremely high accuracy of the Global Positioning System means that we can make maps of trackways accurate to within a few centimeters. New technology is used to document tracks in the field providing accurate information for the creation of 3-dimensional computer images in the laboratory.

The future of dinosaur tracking looks bright. New sites and different track types are being discovered all the time. The opening up of countries like China and Russia, and the expansion of exploration in polar regions, South America, and India means that the variety and quantity of tracks is likely to increase. Even in relatively well-studied areas of North America, new sites are constantly being discovered. One exciting example is the huge Red Gulch Dinosaur Tracksite in Wyoming.[12] These new track discoveries increase our knowledge of the diversity and distribution of dinosaurs and help us gain greater insights into their behavior.

Dinosaur footprints capture the imagination. It's wonderful to think that you can walk on the exact surface where a dinosaur walked one or two hundred million years ago. Tracks are very evocative. As you look along a trackway, you can almost see the animal wandering along ahead of you. Partly because tracks bring dinosaurs to life so convincingly, and partly because they can reveal aspects of behavior of these animals when they were alive, tracks and trackways complement information from body fossils and help us to construct a more complete picture of life in the Mesozoic.

References

1. Chin, K. 1997b. What did dinosaurs eat? Coprolites and other direct evidence of dinosaur diets. In *The complete dinosaur*, eds. J.O. Farlow and M.K. Brett-Surman, 371–382. Bloomington: Indiana University Press.

2. Chin, K. 1997a. Coprolites. In *Encyclopedia of dinosaurs*, eds. P.J. Currie and K. Padian, 147–150. San Diego: Academic Press.

3. Cohen, A., M.G. Lockley, J. Halfpenny, and E. Michel. 1991. Modern vertebrate track taphonomy at Lake Manyara, Tanzania. *Palaios* 6:371–389.

4. Alexander, R.M. 1976. Estimates of speeds of dinosaurs. *Nature* 261:129–130

5. Alexander, R.M. 1989. *Dynamics of dinosaurs and other extinct giants.* New York: Columbia University Press.

6. Farlow, J.O., J.G. Pittman, and J.M. Hawthorne. 1989. *Brontopodus birdi,* Lower Cretaceous sauropod footprints from the U.S. Gulf coastal plain. In *Dinosaur tracks and traces,* eds. D.D. Gillette and M.G. Lockley, 371–394. Cambridge: Cambridge University Press.

7. Lockley, M.G., and A.P. Hunt. 1995. *Dinosaur tracks and other fossil footprints of the western United States.* New York: Columbia University Press.

8. Irby, G.V. 1999. Paleoichnological evidence for running dinosaurs worldwide. *Museum of Northern Arizona Bulletin* 60:109–112.

9. Day, J.J., D.B. Norman, P. Upchurch, and H.P. Powell. 2002. Dinosaur locomotion from a new trackway. *Nature* 415:494–495.

10. Hutchinson, J.R., and M. Garcia. 2002. *Tyrannosaurus* was not a fast runner. *Nature* 415:1018–1021.

11. Thulborn, R.A., and M. Wade. 1979. Dinosaur stampede in the Cretaceous of Queensland. *Lethaia* 12:275–279.

12. Breithaupt, B.H., E.H. Southwell, T.L. Adams, and N.A. Matthews. 2001. Innovative documentation methods in the study of the most extensive dinosaur tracksite in Wyoming. In *Proceedings of the 6th fossil resource conference,* eds. V. Santucci and L. McClelland, 113–122. Geological Resources Division Technical Report.

13a. Lockley, M.G., J.O. Farlow, and C.A. Meyer. 1994. *Brontopodus* and *Parabrontopodus* ichnogen. nov. and the significance of wide- and narrow-gauge sauropod trackways. *Gaia* 10:135–146 (Museu Nacional de Historia Natural, Lisbon, Portugal).

13b. Lockley, M.G., A.P. Hunt, J.J. Moratalla, and M. Matsukawa, M. 1994. Limping dinosaurs? Trackway evidence for abnormal gaits. *Ichnos* 3:193–202.

14. Hitchcock, E. 1836. Ornithlithichnology. *American Journal of Science* 29:307–340.

15. Hitchcock, E. 1858. Ichnology of New England: a report of the sandstone of the Connecticut Valley, especially its fossil footmarks. Boston: William White.

16. Coombs, W.P., Jr. 1980. Swimming ability of carnivorous dinosaurs. *Science* 207:1198–1200.

17. McAllister, J. 1989. Dakota Formation tracks from Kansas: implications for the recognition of subaqueous traces. In *Dinosaur tracks and traces,* eds. D.D. Gillette and M.G. Lockley, 343–348. Cambridge: Cambridge University Press.

18. Ishigaki, S. 1989. Footprints of swimming sauropods from Morocco. In *Dinosaur tracks and traces,* eds. D.D. Gillette and M.G. Lockley, 83–86. Cambridge: Cambridge University Press.

19. Lockley M.G., and A. Rice. 1990. Did *Brontosaurus* ever swim out to sea? Evidence from brontosaur and other dinosaur footprints. *Ichnos* 1:81–90.

20. Lockley, M.G. 1991. *Tracking dinosaurs: a new look at an ancient world.* Cambridge: Cambridge University Press.

21. Gustason, E.R. 2000. *Jurassic highways and byways: a new mechanism for avulsion in the Salt Wash Member of the Upper Jurassic Morrison Formation, east-central Utah.* American Association of Petroleum Geologists Annual Meeting 2000.

22. Kuban, G. 1989. Elongate dinosaur tracks. In *Dinosaur tracks and traces,* eds. D.D. Gillette and M.G. Lockley, 57–72. Cambridge: Cambridge University Press

23. Lockley, M.G., J.L. Wright, S.G. Lucas, and A.P. Hunt. 2001. The late Triassic sauropod track record comes into focus. In *Old legacies and new paradigms,* 181–190. New Mexico Geological Society Guidebook, 52nd Field Conference.

24. Farlow, J.O., and P. Dodson. 1996. Ichnology vs. anatomy? *Ceratopsipes* and the forelimb carriage of ceratopsids. *Journal of Vertebrate Paleontology* 16(3 Suppl.):33.

25. Wright, J.L. 1999. Ichnological evidence for the use of the forelimb in iguanodontid locomotion. *Special Papers in Palaeontology* 60:209–219.

26. Lockley, M.G., J.L. Wright, D. White, J. Li, F.L. Lu, L. Hong, and M. Matsukawa. The first sauropod trackways from China. *Cretaceous Research.* In review.

27. Gatesey, S.M., K.M. Middleton, F.A. Jenkins, and N.H. Shubin. 1999. Three-dimensional preservation of foot movements in Triassic theropod dinosaurs. *Nature* 399:141–4.

28. Thulborn, T. 1990. *Dinosaur tracks.* New York: Chapman and Hall.

Raising The Dead: Excavating Dinosaurs

Daniel J. Chure
Dinosaur National Monument
Vernal, UT

15

Let's get this straight from the outset—dinosaur paleontology is not archaeology. Yes, both archaeologists and dinosaur paleontologists dig things out of the ground, but there is a big difference in the manual labor involved. Archeologists work in soft sediments, using whisk brooms, paint brushes, trowels, and dustpans. For paleontology, forget those tools! Start lifting weights so you can use hammers, chisels, jackhammers, and pneumatic drills for extended periods of time, haul 100-pound bags of plaster up and down hills, and roll over large plaster-encased fossils. For extra fun, drag these heavy loads by hand back to the road where the field truck is parked.

A number of publications discuss in detail how to collect large vertebrate fossils.[1, 2, 3] Here, I will briefly review collecting techniques and take a broader look at many other aspects of digging up dinosaurs. Some aspects have nothing to do with putting a chisel to the rock, but they are nonetheless an important part of the excavation process.

Getting Started

Many factors come into play when deciding where to dig for dinosaurs. Sometimes someone will come to you with a dinosaur bone that they have found. If you are lucky, it will be identifiable as an important specimen and the individual will remember how to get back to the site where the bone was found. That sometimes happens, but not often. Usually where you dig is directed by your research interest.

A professional paleontologist doesn't just walk out into the countryside and look for dinosaurs. There's lots of planning to do beforehand. You need a geologic map of the area to be searched. A geologic map will save time by identifying the rock outcrops most likely to contain dinosaur fossils. There's no point in looking at rocks that are too young or too old. Nor should you waste much time looking in the wrong types of rocks. For instance, while dinosaur remains have occasionally been found in marine rocks, these are very rare occurrences, the result of a dinosaur carcass being carried by a river into

The vast badlands of the western interior of North America are some of the greatest fossil-producing beds in the world, and many areas have been barely explored, if at all.

Dan Chure

Daniel J. Chure received his M.S. and Ph.D. in geology from Columbia University. He has served as paleontologist and research scientist at Dinosaur National Monument for over 20 years. His research interests include the morphology and evolution of theropod dinosaurs, structure and evolution of the Late Jurassic terrestrial ecosystems of North America, and the problem of theft of fossil vertebrates from the field and museums.

the ocean. If you want a dinosaur, it's more productive to look at rocks which were laid down on land.

Next, decide what it is you want to look for. Maybe you are interested in a particular group of dinosaurs. Let's say your specialty is ceratopsians, such as *Triceratops* and its brethren. This group appears relatively late in dinosaur history, during the Cretaceous Period, so don't bother looking at rocks from the Triassic or Jurassic Periods. Alternatively, maybe you

want to fill in gaps in the fossil record. While dinosaurs are well known from the Late Jurassic and Late Cretaceous of North America, those of the Early and Middle Cretaceous are much more poorly understood. Identifying formations of those ages on the map can help focus your efforts within that narrower time frame. Then again, maybe you want to find more of a poorly known dinosaur. For example, there are only a few good specimens of the Late Jurassic meat-eater *Ceratosaurus*. To look for more *Ceratosaurus* fossils, you might decide to search outcrops of the Late Jurassic Morrison Formation in the western United States. Again, the geological map will help you in determining where to focus your efforts.

Paperwork, Paperwork, Paperwork

Long gone are the days when you could simply travel around and collect bones as described by Sternberg.[4, 5] In the modern world you must overcome several hurdles before you can get out to the field to look for fossils. The most important hurdle is getting permission to set foot on the land. Whose land is it? Is it privately owned or owned by the state or federal government? To go out on private land requires the permission of the landowner, preferably in writing so it is clear what you have been authorized to do. The land owner might give you permission to just look and not to collect, or maybe the landowner wants to contribute to the science of paleontology and will let you collect whatever you want (just close the gate behind you so the horses and cows don't get out).

Fig. 1. Although bones close to the surface can be reached with hand tools or jack hammers, sometimes more extreme measures are needed. **A.** Bulldozers at the Cleveland-Lloyd Dinosaur Quarry. **B.** Explosives at Dinosaur National Monument.

One of the great strengths of our political and economic system is the right of private property and private ownership. Landowners are free to dispose of their fossils as they see fit. That right may or may not work to your benefit. Some people just don't want anyone on their land and they don't care if the dinosaur fossils turn to dust. Others may let you dig for years and collect a scientific treasure trove. Some may lease their land to commercial dealers for a share of the profits from any fossils found and sold. On rare occasions, paleontologists have worked at a site for some time only to find out the next season that they are denied access and commercial fossil dealers were collecting and selling the bones. Such an event might be maddening, but there is no law that can protect your find, even if the bones are be-

ing ground up for road fill. It's the price we pay for living in a free and open society.

Commercial collecting and selling of vertebrate fossils is prohibited or very restricted on state and federal lands. In virtually all cases, scientific collecting on public lands requires one or more permits. It could take months to process, review, and approve permit applications. There are often land use and resource issues that need to be considered. Are there any threatened or endangered species in the area? Are there archeological or other cultural resources near the site? Is the site located in a Wilderness Study Area or a designated Wilderness area? If so, it may not be possible to get a permit or there may be burdensome restrictions accompanying the permit.

Getting to the Fossils (Almost)

Assuming that the permit hurdles have been navigated, the site itself may still pose some challenges. If you are lucky, the fossil-producing rock layer is horizontal and exposed at the surface. That will make excavation relatively simple. If the layer is not at the surface, remov-

Fig. 2. The removal of the bones from the enclosing rock may destroy paleontological and geological data; so careful recording of observations is crucial. Snappy hat and pipe are optional.

Fig. 3. Each bone must be drawn on a map that will serve as the permanent record of how the bones were distributed in the quarry.

Fig. 4.

Each bone in the quarry receives a unique reference number that is recorded on the quarry map. Here a left premaxilla of an *Allosaurus* proudly bears the field number 445.

small, dinosaur skeletons contain more than 200 bones each. Unless you are very, very lucky, these bones will be partly to completely scattered in the quarry. In most cases, your site will include the bones of several to many skeletons, not necessarily all of the same species. Careful mapping of the position of the bones is important for later research—unraveling what bones belong to which species and which individuals (Fig. 3). Quarry mapping is most often done with a grid system, paper, and drawing tools. Each bone is numbered and recorded on the map (Fig. 4). When bones or groups of bones are put in plaster jackets for removal, each jacket also gets a number that identifies which bones are inside. These records are helpful in determining priorities back in the lab. The information recorded in these maps may also help answer important questions about how the fossil deposit was formed, how far specimens were transported in a river, the disarticulation of skeletons by scavenging, etc.[6] Even today, paleontologists use maps of dinosaur quarries excavated in the 19th century.[7, 8]

Recently, new and more sophisticated technology has come into use. Tethered blimps with cameras have been used in mapping large fossil localities.[9] Instruments called gamma scintillation detectors have been used to locate bones still buried[10, 11] (Fig. 5) and ground penetrating radar has been used to excavate a *Seismosaurus* skeleton.[12] Global positioning systems can help pinpoint a locality so that future researchers can relocate it. Additional documentation of the fossils and the excavation process can be done with videotape (see Chapman, page 137).

Some important kinds of geological data are likely to be destroyed as the bones are excavated. These data will need to be recorded. Valuable information includes the type of rock,

ing the overburden (the rock layers above the fossils) can be difficult work depending on how much has to be removed and how hard the rock is (Fig. 1). This step can involve jackhammers, bulldozers, explosives, or sometimes all three. In large, fossil-rich quarries, overburden will need to be removed over a period of years in order to continue the excavations. Obviously, all this work needs to be done carefully so as not to damage the fossils underneath (Fig. 2).

Once the dinosaur bones are revealed, you will need to record data about their geological context before and during the digging. What data to record and how to do it should be decided at the start. Beginning half way through the excavation will reduce the value of the data.

Foremost in importance are the data regarding the position of the bones. Large or

Fig. 5. Ray Jones and his gamma scintillator. The cylindrical sensor he is holding detects gamma emissions from bones, which, under the right conditions, can detect fossils up to three feet below the surface.

records to be used by researchers working on the specimens.

Excavating the Fossils

The tools we use for excavation depend, to a large extent, on the kind of rock in which the fossils occur—matrix. Sometimes, such as in the Cretaceous beds of Mongolia, the matrix is a relatively soft sandstone, which can be readily removed using hand tools. With this type of rock much, but not all, of the rock is removed in the field before jacketing the fossil in plaster (Fig. 6). This soft sandstone can be easily re-

Fig. 6.
A. The bone is exposed and consolidated.
B. A plaster jacket can be applied to protect it during transport.

any sedimentary structures present, the size of the grains in the sandstones, the extent of the bone-bearing layer and its relationship to the beds above and below. This information may help you understand the factors responsible for forming the deposit, such as water current direction. During the dinosaur gold rush of the 19th and early 20th centuries such data were rarely collected. In many cases, even the bone maps were not made. Those losses cannot be regained, leaving us with scientific questions that will probably never be answered.

All the data recorded in field notes, as well as photo and video documentation, are placed in the archives of the museum which will store and care for the fossils. The data will become a part of the museum's official collection

Fig. 7. An *Allosaurus* skeleton, partly exposed and partly jacketed, is being excavated at Dinosaur National Monument by Rod Joblove and Scott Madsen. This specimen is tilted at 70 degrees (110 degrees from horizontal).

Here we use tougher tools—rocksaws, small pneumatic powered jackhammers, as well as regular hammers and chisels. When the rock is hard, we can only rough out the skeleton, leaving a considerable amount of rock with the fossil in the jacket (Fig. 7). This makes the bundle quite heavy (Fig. 8). Heavy fossil jackets are by no means unique to Dinosaur National Monument. Field crews for the famous 19th century Yale paleontologist O.C. Marsh collected and shipped a jacketed *Triceratops* skull that weighed 6,000 pounds!

As you uncover dinosaur bones, there will be small (or large) cracks running through them. Bones may be crushed, slightly or badly distorted, shattered, or partly turned to powder. You will need to apply various types of glues and hardeners to the bones and matrix, in order to stabilize the fragments in place until you can piece them back together in the lab (Fig. 9). Hardeners and glues are usually applied with a brush. Hardeners are dissolved in acetone, which rapidly evaporates, so they dry quickly. Several coats may be needed to stabilize a badly damaged specimen.

Fig. 8. The *Allosaurus* skeleton completely jacketed, but still in place. This 5700-pound block posed a challenge because it had to be lowered backwards (towards the right side of the photograph). The timbers will form the pallet for the jacket shown in the subsequent figures. Ron Hopwood (left) and Rod Joblove (right) ponder the work ahead.

moved back in the laboratory, exposing bones as though they had come from a just-deceased animal.[13,14]

In other cases, such as at Dinosaur National Monument, the sandstone is quite hard.

Fig. 9.

Fossil dinosaur bones are fragile and need to be strengthened and consolidated prior to jacketing. Acetone-based consolidants dry quickly and help protect the fossil during excavation and transport.

Once stabilized, the process of getting the blasted thing out of the ground begins. We start by digging a trench around the specimen, down to a level safely below the bottom of the fossil. Even well fossilized bone is brittle and can be broken from vibrations during transport, or from being bumped or dropped. So you need to protect the bone during transport from the field to the lab. The most common technique

Fig. 10. A. Distant view of the *Allosaurus* quarry with the pallet in place. **B.** Close-up showing Ron Hopwood (left) and Rod Joblove (right) using nylon webbing to strap the jacket to the pallet.

involves using burlap strips dipped into plaster and applied around the block of fossils. However, the plaster burlap by itself will stick to the fossil if applied directly, making it impossible to remove back in the lab without damaging the bone. So a separator, usually just toilet paper, is first applied to the rock and fossils. It is laid across the block several layers thick, then moistened with water and a brush. The water makes the separator conform more closely to the irregular rock surface and makes a protective jacket. Next, burlap strips are dipped in plaster and applied in several crisscrossed layers around the top and sides of the block. If the block is large and heavy, wooden boards or metal rods can be added to give the jacket extra strength.

Remember when we dug your trench around the fossil block, digging to a depth below the fossil? Now get back into the trench and start digging horizontally under the block. As you dig, prop the block up as needed and periodically apply toilet tissue and plastered burlap to the newly exposed underside of the block. Ideally, you will end up with the jacketed fossil block supported on a narrow pedestal. At this point the jacket is popped off its pedestal, usually using a rod as a lever to pry it

off. And with this, the moment of truth has arrived. If the jacket fits well and the rock and bone inside have been well consolidated, everything should go well. Everything stays put inside the protective jacket. However, sometimes softer sedimentary rock, such as marine shale, is badly fractured, and when the jacket is popped off, the contents pour out the unplastered portion on the bottom of the block. UGH—a real Kodak moment!

Getting the Bones Back to the Lab

When our excavation is finished for the season, you will have accumulated some number of plaster jackets. In many cases the jackets are too heavy to carry out in your backpack. So what do you do?

If you can drive a vehicle to the quarry site, hauling, sliding, and pushing the jackets onto the truckbed may not be too difficult. Jim Jensen of Brigham Young University mounted

Fig. 11. Ron Hopwood, who designed the pallet, examines his masterpiece after the pallet and jacket are pried loose and lowered backwards by ropes. The entire package was lifted by a helicopter back to the lab.

Fig. 12.

The jacketed *Allosaurus* skeleton back at the lab after its helicopter ride. The smiling crewmembers are (right to left) Marilyn Sokolowski, Rod Joblove, Scott Madsen, and Ann Elder.

ronmental damage than simply driving a vehicle to the site.

Sooner or later, you will have blocks that are just too big to move by normal means (Fig. 10). In such cases, large helicopters can come in, lift the jackets, and carry them to your vehicle (Fig. 11). However, helicopter rental time is expensive. Sometimes helicopter companies will donate their time. In special cases, the Air National Guard, or another branch of the U.S. military, might help out and provide a helicopter to move the fossils (Fig. 12).

a steel tripod with a block and tackle onto the back of a large truck to help lift heavy blocks, some weighing several tons.

On the other hand, the quarry may be in a place to which a vehicle cannot be driven, or in an area where driving off-road is prohibited. If the jackets aren't too heavy, they may be loaded on skids and dragged by hand or horse back to the road. However, this may cause more envi-

Tidying Up

If the quarry will take several years to excavate, you will not get all the jackets out in one season. That means covering them with cloth tarps or plastic sheeting before you leave and backfilling dirt over them. Yes, that means re-excavation at the beginning of the next field season, but at least they will be protected from rain or snow damage and from possible vandalism and theft. Even though we have been excavating the land, it is important to leave the site looking as undisturbed as possible.

And Finally . . .

For many, part of the allure of digging dinosaurs is traveling to foreign lands, meeting exotic peoples and cultures, and roaming the wilds abroad in search of new and wondrous beasts. However, all the difficulties discussed above are greatly magnified in an overseas expedition. Paperwork (permits, visas, passports, etc.) is much more compli-

Fig. 13. Worth the effort. Back in the lab the jacket is opened and the spectacular *Allosaurus* skeleton is prepared. Not shown: (a) the neck, which was in a separate jacket and (b) the skull, which was found by Ray Jones using a gamma scintillator, two years after the skeleton was removed and the quarry abandoned.

cated. Weeks can be lost sitting at a border trying to get into the country and shipping trucks, fuel, and supplies across the globe can be time-consuming and expensive.[13] And, having a vehicle break down in the Sahara Desert is a tad more inconvenient than having that happen in rural Montana.

To be realistic, the world is far from a safe place and there are places where strangers are not welcomed. Political chaos, civil wars, and ethnic and religious conflicts can make an overseas expedition perilous. Strangers in remote areas with new vehicles, electronic equipment, and valuable supplies could be a target or may not be well received. These problems are not new. The American Museum of Natural History's Central Asiatic Expeditions in the 1920s had problems with bandits, as well as with combatants in the ongoing Chinese civil war. Many countries however, including the United States, continue to welcome foreign scientists. Such international collaborations have led to exciting scientific discoveries[13, 14] and at the same time may help foster understanding and respect among different cultures.

Needless to say, you do not have to go around the world to discover dinosaurs. The vast badlands of the Western Interior of North America are some of the greatest fossil-producing beds in the world, and many areas have been barely explored, if at all. New discoveries continue to be made in this region and will continue far into the future. Regardless of where you go, good luck and happy dinosaur hunting (Fig. 13).

References

1. Chure, D.J. 2000. Digging them up. In *The Scientific American book of dinosaurs*, ed. G. Paul, 46-51. New York: St Martin's Press.

2. Hotton, N. 1965. Tetrapods. In *Handbook of paleontological techniques*, eds. B. Kummel and D. Raup, 119–125. San Francisco: W.H. Freeman and Co.

3. Leiggi, P., C.R. Schaff, and P. May. 1994. Macrovertebrate collecting: field organization and specimen collecting. In *Vertebrate paleontological techniques*, Vol. 1, eds. P. Leiggi and P. May, 59–77. New York: Cambridge University Press.

4. Sternberg, C.H. 1909. *The life of a fossil hunter*. Henry Holt and Company (Reprinted 1990, Indiana University Press, Bloomington).

5. Sternberg, C.H. 1917. *Hunting dinosaurs in the bad lands of the Red Deer River, Alberta, Canada*. Lawrence, Kansas: Published by the author (Reprinted 1985 NeWest Press).

6. Riggs, R.R. 1994. Collecting taphonomic data from vertebrate localities. In *Vertebrate paleontological techniques*, Vol. 1, eds. P. Leiggi and P. May, 47–58. New York: Cambridge University Press.

7. Monaco. P. 1998. A short history of dinosaur collecting in the Garden Park Fossil Area, Canon City, Colorado. *Modern Geology* 234:465–480.

8. McIntosh, J.S. 1998. New information about the Cope collection of sauropods from Garden Park, Colorado. *Modern Geology* 23:481–506.

9. Breithaupt, B.H., and N.A. Matthews. 2001. Preserving paleontological resources using photogrammetry and geographic information systems. In *Crossing boundaries in park management*, ed. D. Harmon, 62–70. The George Wright Society Biennial Conference, Denver, CO.

10. Jones, R., G.H. McDonald, and D.J. Chure. 1998. Using radiological surveying instruments to locate subsurface fossil vertebrate remains. In *Partners preserving our past, preserving our future*, eds. J.E. Martin, J.W. Hogenson, and R. Benton. *Dakoterra* 5:84-90 (Museum of Geology, South Dakota School of Mines).

11. Jones, R., and D.J. Chure. 2000. The recapitation of a Late Jurassic theropod dinosaur: successful application of radiological surveying for locating subsurface fossilized bone. In *Special volume on aspects of theropod paleobiology*, eds. B.P. Perez-Moreno, T. Holtz, J.L. Sanz, and J. Moratalla. *Gaia* 15:103–110 (Museu Nacional de Historia Natural, Universidade de Lisboa, Portugal).

12. Gillette, D.D. 1994. *Seismosaurus: the Earth shaker*. New York: Columbia University Press.

13. Novacek, M. 1996. *Dinosaurs of the Flaming Cliffs*. New York: Anchor Doubleday Books.

14. Clarke, J.M., M.A. Norell, and R. Barsbold. 2001. Two new oviraptorids (Theropoda:Oviraptorosauria), Upper Cretaceous Djadkhta Formation, Ukhaa Tolgod, Mongolia. *Journal of Vertebrate Paleontology* (21) 2:209–213.

A c k n o w l e d g m e n t s

I thank the three anonymous reviewers for their constructive comments that improved this paper. Figs. 1A and 2–6 courtesy of James H. Madsen Jr., DINOLAB. All other photos courtesy of the Department of the Interior/National Park Service.

Applying 21st Century Technology to Very Old Animals

16

Ralph E. Chapman, *Smithsonian Institution, Washington, DC*
Neffra A. Matthews, *Bureau of Land Management, Denver*
Mary H. Schweitzer, *Montana State University, Bozeman*
Celeste C. Horner, *Montana State University, Bozeman*

In one of the first scenes of the widely successful movie *Jurassic Park*, a vertebrate paleontologist stands in a tent with others in his field party watching a computer screen. An explosive charge is set off and the three-dimensional image of a dinosaur skeleton, still buried in the rock of the outcrop appears on the screen. Crichton describes this technique in his novel as Computer-Assisted Sonic Tomography (CAST), a term he invented. The rest of the novel, and the movie based on it, explodes with the glories—and dangers—of mixing dinosaurs with technology, culminating with the cloning of dinosaurs.[1] This vision is far from the truth. Despite tremendous advances made in using technology to find and study dinosaurs, we are not yet at the point where we can do such a fine job of visualizing a dinosaur still in the rock, although we are getting closer. Likewise, our ability to clone dinosaurs is non-existent, at least for extinct dinosaurs (but not for some birds). Happily, the dangers of mixing technology with dinosaurs seem to be relatively minimal.

It is essential that paleontologists continue to use technology extensively in their work with dinosaurs and other extinct vertebrates.[2] Why is this so? Paleontology is a very difficult science to pursue in any quantitative or rigorous way because it is an historical science complicated by the great variability typical of biological systems. Further, significant difficulties are introduced by the incomplete nature of the fossil record and the non-random way it can be sampled by paleontologists. Vertebrate paleontologists, especially those who study dinosaurs, suffer the extreme of these problems—typified by a relatively limited area to explore for fossils and small sample sizes for their organisms. Many genera are represented by only a single, partial specimen. This difficulty can be compounded by other problems such as the fragility of the fossils and the enormous size of some of the animals. This situation makes it vital that the maximum amount of information be extracted from those samples and fossils we do have.

> *Technology will not only make the study of dinosaurs progress much faster, it will allow us to see into the past in ways never before possible.*
>
> Chapman, Matthews, Schweitzer & Horner

Ralph E. Chapman is a paleontologist who runs the Applied Morphometrics Laboratory in the National Museum of Natural History, Smithsonian Institution. He has degrees in zoology from the University of Bridgeport, CT, and in paleontology from the University of Rochester, NY. His research projects include paleontology, forensics, anthropology and archaeology, functional morphology, and systematic biology. Chapman has done work on dinosaurs, trilobites, fossil humans and a host of other organisms.

Neffra A. Matthews, Chief of the Branch of Photogrammetric Application at the Bureau of Land Management's National Science and Technology Center in Denver, received B.S. and M.S. degrees in geology from the University of Kentucky. As a photogrammetist, she has produced many maps

continued on p. 139

137

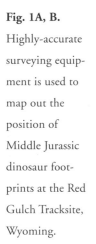

Fortunately, technology can do a lot to help us study dinosaurs. We will discuss the work done by paleontologists in finding, extracting, and studying dinosaurs, and how technology has, is currently, and will be helping with this process. We will start, as paleontologists do, in the field. There, technology helps us figure out where we are in the world and in the time column, determine where the fossils may be, map where they are in the ground, and collect the important information

was found in the modern world and where that place was on Earth back when the animal lived and died. To further complicate things, the fossil may have been transported after the animal died, sometimes thousands of years or more later, and the animal itself may have migrated

Fig. 1A, B.

Highly-accurate surveying equipment is used to map out the position of Middle Jurassic dinosaur footprints at the Red Gulch Tracksite, Wyoming.

available from the context of where they were found. Technology also can provide significant input into the efficient removal and preparation of the fossil material. Next, we will discuss how technology helps us study fossils and the organisms that produced them. Many areas of research have been revolutionized by the technology of faster computers and better analytical methods, but we will concentrate here on the technological part of getting the original data. This chapter will partially concentrate on scanning and related methods that allow the more exact reconstruction of the animals and a better understanding of how they were preserved and some aspects of how they lived. We also will discuss how we can extract biochemical information from fossils and what the potentials are for extracting ancient DNA from a select few of them.

Getting the Fossils and Data

Dinosaurs and other extinct forms have a very complex relationship with geography. Every specimen is characterized by the place where it

hundreds of miles throughout its life. It takes considerable work and analysis to dissect all these components and suggest the real story for the animal that produced the fossil and the species to which it belongs. Technology not only can help us do the analyses—it can also help us find and preserve the clues to this history that are present in the context of where the fossil was recovered.

We will start with the location where the fossils are found, defined by the modern longitude and latitude. Technology can provide considerable help here, and the way paleontologists map where they are in the field has totally changed over the past decade or so. We now use Global Positioning Systems (GPS), receivers that pick up coded signals from navigational satellites. When signals are available from four or more of these satellites, a position can be determined mathematically with an accuracy depending on how precise the GPS clock is, how strong the signals are, and how long the unit can communicate with the satellites. The result is a coordinate position with resolution varying from tens of meters to just a few centimeters. GPS systems are not only used to identify where fossils come from, but also are a great help in finding and relocating fossil localities that are expected to produce new material or have produced important material in the past. When the American Museum of Natural History started their newest explorations of the Gobi Desert, they used GPS systems to relocate the old localities made famous by Roy Chapman Andrews and to find their way around parts of Mongolia where quality maps were unavailable.[3]

Finding fossils in the field can be a real art. Like many artistic disciplines, prospecting can be done much better using a solid scientific background and with selected technology. For more than a century, paleontologists have been using geological maps to find areas that contain the right kinds of rocks (from the right environments) of the right age (Late Triassic to Late Cretaceous) for dinosaurs. This research can be done with far greater power using Geographic Information Systems (GIS), complex databases that take large amounts of paleoenvironmental and geological data distributed across broad geographic areas, determine critical combinations of these factors, and map out those places where paleontologists have the greatest chance of finding the fossils they are seeking.

On the more local level, prospecting has been attempted using various "scanning" technologies along the lines of the CAST method envisioned by Crichton in *Jurassic Park*. The methods used include remote-sensing approaches, geophysical diffraction tomography, ground-penetrating radar, scintillation counting, and magnetometry. Paleontologist David Gillette has investigated these methods the most, working on the skeleton of the huge sauropod *Seismosaurus* in the field in New Mexico in cooperation with scientists from the Los Alamos, Oak Ridge, and Sandia National Laboratories.[4, 5] So far the results are mixed because many bones are difficult to distinguish from the rocks that contain them. This approach should be an area of significant development in coming years, however. In special cases, dinosaur bones have incorporated radioactive material during the process of fossilization and can be hunted using Geiger counters.

Quarrying the Quarry

Once we find an important locality, technology can be an enormous help in mapping the fossils and recording the important contextual data. This step is very important because a great amount of information is available from how the fossils and sediments are distributed, including the other animals and plants that lived in the area. These data can indicate the environment that produced the outcrop and fossils and even significant bits about the biology of the dinosaurs found there. A great limitation of many of the dinosaur fossils found over 50 years ago is that much of this information was never recorded; unfortunately, much of it still is not collected in many cases. Electronic Distance Measurement devices (EDMs) and other advanced surveying equipment can help locate fossils within a quarry with sub-centimeter accuracy. These units use lasers and other optical systems and provide spatial coordinates for the fossils relative to known benchmarks (Fig. 1). For elongate fossils, the position of each end is recorded to provide orientation data. In

continued from p. 137

from aerial photographs. Recently, she has been using close-range photogrammetric techniques to document paleontological resources on public lands in the West. Before coming to BLM, she worked as a geologist for the Defense Mapping Agency.

Mary H. Schweitzer is Adjunct Assistant Curator of Paleontology at the Museum of the Rockies, Adjunct Assistant Research Professor in Biology and Earth Science, and Assistant Research Professor in Microbiology at Montana State University. She is one of the leading authorities on ancient biomolecules in vertebrate paleontology.

Celeste C. Horner is the Computer Specialist in the Computational Paleontology Lab at the Museum of the Rockies at Montana State University in Bozeman, Montana. She has worked on many museum projects applying state-of-the-art techniques to various applications in vertebrate paleontology, e.g., CT-scan analysis, morphometrics, 2-D and 3-D specimen reconstruction, and biomechanical simulations.

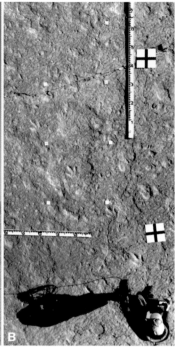

Fig. 2.
Detailed mapping effort for Middle Jurassic Red Gulch Tracksite, Wyoming. **A.** Blimp used to capture high-level view of entire tracksite. **B.** Photograph taken of a large section of the tracksite using the blimp. **C.** Photographic recording of smaller, individual grids within tracksite.

some cases, blimps or other unmanned flyers (Fig. 2) can be useful in mapping quarries.[6, 7] Similar methods can place the fossils more accurately within the time column—when the animals lived—and the technology of radiometric dating can now provide solid dates for some associated sediments.

GIS systems, with their advanced graphical capabilities, can allow paleontologists to construct a virtual map of a quarry and all the fossils and other elements found within it. The virtual map can be used not only as a primary research tool, but for exhibition and educational applications as well; individuals can tour through each quarry virtually and select individual components to see even more detailed data including high resolution images and three-dimensional models.

Being Prepared

Technological advances have yet to be applied to fossil preparation to any significant degree, but they should provide great advancements over the next decade or so. Long preparation time for the specimens is one of the great problems in vertebrate paleontology; specimens are removed from the field in huge plaster jackets with only partial data on where the fossils may

be within them. In the preparation laboratory, much time is spent figuring out what is really contained in these blocks. Assistance in knowing where the important material is can cut preparation time considerably. Many of the same methods discussed earlier for finding the bones in the ground can be combined with other scanning technologies, such as CT-scanning, to help guide the paleontologist preparing a specimen once it gets back to the laboratory by allowing a peek at what is inside the jacket before preparation.

Making Virtual Dinosaurs

Fossils can be difficult to work with when trying to reconstruct dinosaurs. Many elements can be missing, they are often distorted and/or damaged, they can be either too large or too small to handle, and they typically can be very heavy and yet very fragile. Some fossils, including most footprints, are difficult to remove from the field. Technology can help considerably by allowing us to scan the bones or traces in three-dimensions and then rebuild the whole fossil virtually in the computer.[8, 9]

Most people are familiar with standard, medical Computerized Tomographic (CT) scanning systems, and these are now used extensively by paleontologists (Fig. 3A). Higher-resolution systems are available for smaller specimens. The data are studied as in-

dividual slices through the fossil, or assembled into full three-dimensional objects (Fig. 3B). This scanning makes it possible to isolate structures, even on the inside of the specimen, such as the braincase or embryonic bones within eggs.[10, 11]

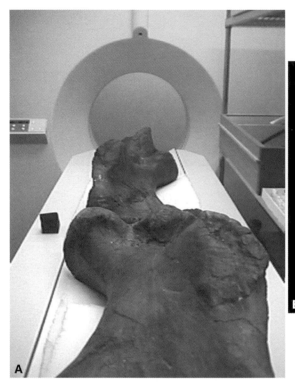

Three-dimensional surface scanners (Fig. 4A), most using lasers or other optical methods, provide very high-resolution data for objects ranging in size from microchips to buildings (Fig. 3B). Finally, photogrammetry, the process of making measurements using photographs, also can generate similar three-dimensional data and this method excels at being used directly in the field with dinosaur trackways and other objects that cannot be removed.[6, 7]

All of these systems produce three-dimensional reconstructions of specimens in the computer. These virtual fossils can be studied alone or assembled into virtual skeletons which can be animated and/or studied biomechani-

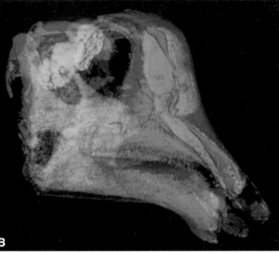

cally (Fig. 4B). For example, Rayfield and colleagues were able to distinguish between meat-slashing and bone-crushing predators by applying engineering software to CT-scans of skulls.[12] The virtual fossils also can be modified to take away the distortion caused by the process of fossilization, repaired if broken, have missing parts replaced, scaled to different sizes, and used as a basis for accurately fleshing-out reconstructions of extinct animals.[13]

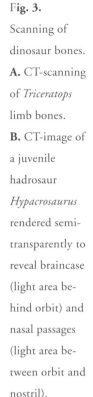

Fig. 3. Scanning of dinosaur bones. **A.** CT-scanning of *Triceratops* limb bones. **B.** CT-image of a juvenile hadrosaur *Hypacrosaurus* rendered semi-transparently to reveal braincase (light area behind orbit) and nasal passages (light area between orbit and nostril).

Fig. 4. Laser scanning of the Smithsonian *Triceratops*. **A.** Laser scanning of the back of the frill (skull). **B.** Complete virtual *Triceratops* built using surface scanning.

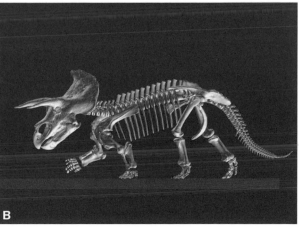

Another technology called rapid prototyping can produce three-dimensional, highly-accurate replicas (prototypes) of these virtual specimens.[14] These replicas can be produced by systems that carve the model out of a block of composite material. More accurate systems build the prototypes layer by layer using either sliced sheets of paper fused with adhesive or by depositing or solidifying a liquid to form each thin layer using lasers or some other technology. The most common method is stereolithography. It uses a laser to solidify each thin layer of a photo-curable polymer to build exquisitely accurate models of the original shape. Prototyping is now used extensively by vertebrate paleontologists and can produce accurate replicas at the same scale as the original fossil, at a very reduced size (Fig. 5A), or significantly enlarged (Fig. 5B). Prototyping also can allow the accurate generation of missing material by making mirror-images of other bones, scaling up or down repeated elements such as ribs or vertebrae, or by assembling composite bones out of the good parts of multiple specimens.

Blood from a Stone

Much has been written about the possibility of extracting the DNA from dinosaur fossils since well before the publication of *Jurassic Park*. The possibility of cloning a dinosaur and bringing these marvelous beasts "back to life" has caught the imagination of scientists and the general public alike. However, a whole host of problems probably will forever hinder the resurrection of a non-avian dinosaur in this way. First of all, to clone a dinosaur will require the complete component of the DNA for that animal. At the present time, not even a small fragment of DNA has been verifiably recovered from a dinosaur bone.[15] Also, we have no idea how many base pairs would constitute the whole DNA for a single species. If we did have significant amounts of dinosaur DNA from a single animal, additional material from birds or crocodiles—groups closely related to dinosaurs—might be used to help complete the DNA component. Frogs, as used in *Jurassic Park*, are far too distantly related to dinosaurs to be of much use. The results would end up

Fig. 5. Prototyping of the Smithsonian *Triceratops*. **A.** Small sculpture (1/6 scale) of the whole skeleton in old mount posture done using casts of the miniature prototype bones. This sculpture was useful to paleontologists reconstructing the posture of the animal; the miniature bones were far easier to manipulate than the real-sized bones. **B.** Enlarged skull of the Smithsonian *Triceratops*. Skull is assembled from more than 30 blocks of stereolithography fused together and represents a 15 percent enlargement of the original six-foot-long skull. The original skull was determined to be too small for the rest of the skeleton; the original mount was a composite of more than ten individuals.

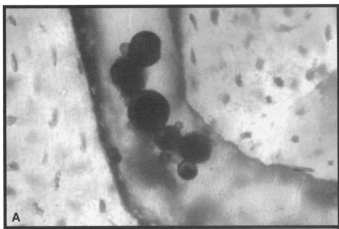

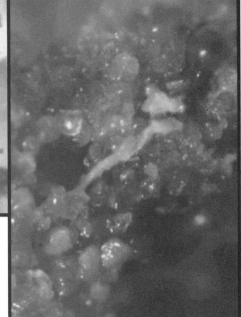

Fig. 6.
A. *Tyrannosaurus rex* vascular microstructures.
B. Epidermal fibers from dinosaur *Shuvuuia deserti*.

being a chimera rather than a real, resurrected dinosaur. However, the total lack of our knowledge in how the DNA of a dinosaur is organized, combined with many other unknowns related to the non-genetic aspects of development and how to keep a dinosaur alive once hatched probably makes cloning an impossibility.

While there is great skepticism and no direct evidence that DNA can last sufficiently long to be preserved from the Mesozoic, other biomolecules can be detected in fossil organisms, including dinosaurs.[15] These biomolecules can provide valuable information about the original animal and the environment in which it lived. This information includes finding molecular signals consistent with collagen, as well as the protein hemoglobin, and the possible remains of preserved red blood cells from a bone of *Tyrannosaurus rex* (Fig. 6A).[16, 17, 18] Additionally, molecular signatures for a specific type of keratin, the protein that makes up bird feathers, have been detected from fossil epidermal structures of the dinosaur *Shuvuuia* (Fig. 6B) and the ancient primitive bird *Rahonavis*.[19, 20, 21] This gives implications for the origin of feathers, as well as more evidence for the close relationship between some

dinosaurs and birds. Clearly, as our technology improves, especially our biotechnology for these studies, we expect other surprising and major breakthroughs over the next decade that should allow us to learn even more about dinosaur biology.

Conclusions

Mixing dinosaurs with technology is not dangerous, it is essential. Technology will be a major part of our effort to expand significantly our knowledge of what these animals were like—how they looked, how they lived—and to continue to find more and more specimens in the field while retaining the essential information that is available only at the outcrop. Technology will not only make the study of dinosaurs progress much faster, it will allow us to see into the past in ways never before possible.

References

1. Crichton, M. 1990. *Jurassic park*. New York: Ballantine.

2. Chapman, R.E. 1997. Technology and the study of dinosaurs. In *The complete dinosaur*, eds. J.O. Farlow and M.K. Brett-Surman, 112-135. Bloomington: Indiana University Press.

3. McKenna, P.C. 1992. GPS in the Gobi: dinosaurs among the dunes. *GPS World* 3(6):20–26.

4. Gillette, D.D. 1994. *Seismosaurus: the Earth shaker*. New York: Columbia University Press.

5. Witten, A., D.D. Gillette, J. Sypniewski, and W.C. King. 1992. Geophysical diffraction tomography at a dinosaur site. *Geophysics* 57:187–195.

6. Matthews, N.A., and B.H. Breithaupt. 2001. Close-range photogrammetric experiments at Dinosaur Ridge. *The Mountain Geologist* 38(3):147–153.

7. Breithaupt, B.H., E.H. Southwell, T.L. Adams, and N.A. Matthews. 2001. Innovative documentation methodologies in the study of the most extensive dinosaur tracksite in Wyoming. In *Proceedings of the 6th fossil resources conference*, eds. V.L. Santucci and L. McClelland, 113–122. US DOI, National Parks Service, Geological Resources Division Technical Report NPS/NRGRD/GRDTR-01/01.

8. Chapman, R.E., G. Hunt, and D. Rasskin-Gutman. 1996. Three-dimensional scanning, digitization and modeling of dinosaur fossils. *Journal of Vertebrate Paleontology* 16(3, suppl.):27A.

9. Horner, C.J., J. Hutchinson, and E. Lamm. 1998. 3-Dimensional imaging and reconstruction: an overview of techniques and results. *Journal of Vertebrate Paleontology* 18(3, suppl.):52A.

10. Brochu, C.A. 2000. A digitally-rendered endocast for *Tyrannosaurus rex*. *Journal of Vertebrate Paleontology* 20(1):1–6.

11. Horner, J.R. 1997. Rare preservation of an incompletely ossified embryo. *Journal of Vertebrate Paleontology* 17:431–434.

12. Rayfield, E.J., D.B. Norman, C.C. Horner, J.R. Horner, P.M. Smith, and J.J. Smith. 2001. Cranial design and function in a large theropod dinosaur. *Nature* 409:1033–1037.

13. Walters, R., R.E. Chapman, and R.A. Snyder. 2001. Fleshing-out *Triceratops*: adding muscle and skin to the virtual *Triceratops*. *Journal of Vertebrate Paleontology* 21(3, suppl.):111A.

14. Andersen, A., R.E. Chapman, J. Dickman, and K. Hand. 2001. Using rapid prototyping technology in vertebrate paleontology. *Journal of Vertebrate Paleontology* 21(3, suppl.):28A.

15. Hedges, S.B., and M.H. Schweitzer. 1995. Detecting dinosaur DNA. *Science* 268:1191–1192.

16. Schweitzer, M.H., M. Marshall, K. Carron, D.S. Bohle, S. Busse, E. Arnold, C. Johnson, and J.R. Starkey. 1997A. Heme compounds in dinosaur trabecular tissues. *Proceedings of the National Academy of Sciences* 94:6291–6296.

17. Schweitzer, M.H., C. Johnson, T.G. Zocco, J.R. Horner, and J.R. Starkey. 1997B. Preservation of biomolecules in cancellous bone of *Tyrannosaurus rex*. *Journal of Vertebrate Paleontology* 17(2):349–359.

18. Schweitzer, M.H., and J.R. Horner. 1999. Intravascular microstructures in trabecular bone tissues of *Tyrannosaurus rex*. *Annals de Paleontologie* 85(3):179–192.

19. Schweitzer, M.H., J.A. Watt, R. Avci, L. Knapp, L. Chiappe, M. Norell, and M. Marshall. 1999A. Beta keratin specific immunological reactivity in feather-like structures of the Cretaceous alvarezsaurid *Shuvuuia deserti*. *Journal of Experimental Zoology* 285:146–157.

20. Schweitzer, M.H., J.A. Watt, R. Avci, C.A. Forster, D.W. Krause, L. Knapp, R.R. Rogers, I. Beech, and M. Marshall. 1999B. Keratin immunoreactivity in the Late Cretaceous bird, *Rahonavis ostromi*. *Journal of Vertebrate Paleontology* 19(4):712–722.

21. Schweitzer, M.H. 2001. Evolutionary implications of possible protofeather structures associated with a specimen of *Shuvuuia deserti*. In *New perspectives on the origin and early evolution of birds: proceedings of the international symposium in honor of John H. Ostrom*, eds. J. Gauthier and L.F. Gall, 181–192. New Haven: Yale University Press.

A c k n o w l e d g m e n t s

We would like to thank a number of individuals for their help in preparing this chapter. Linda Deck and Rebecca Snyder helped directly with the assembly of the manuscript and provided essential editing. We also thank Art Andersen, Lisa Federici, Jason Dickman, Ron Jones, Bob Walters, Jack Horner, Brent Breithaupt, Judy Scotchmoor, and the whole Smithsonian *Triceratops* team, especially Richard Benson and Steve Jabo. We would also like to thank our Institutions for their support of our work; the Smithsonian Institution, the Bureau of Land Management, the Museum of the Rockies, and the Montana State University.

Depicting Dinosaurs: How We Build, View, and Portray Them

Ever wondered what it takes to go from the bones in the rocks to the scaly, toothy dinosaur in a painting, sculpture, or Hollywood epic? The article by **Walters and Kissinger** will show you how it is done. These "paleo artists" explain how they work with paleontologists to put the flesh on the bones of long-dead dinosaurs, and to recreate as accurately as possible the scenes of Mesozoic Earth that are so familiar to us today.

Dodson tackles the questions kids ask and adults can't answer. No, this article isn't about embarrassing topics! It's about the first, the biggest, the smallest, the brainiest, the dumbest, the fastest, the longest, the "most-est" among the dinosaurs. Want to know which dinosaur laid the biggest egg, or why North America has so many dinosaur fossils? Here are your answers....

Megalosaurus to modern media monsters, **Springer and Platt** take readers on a journey through the history of dinosaur illustration. They explain how dinosaurs became "stars" with the help of artists and the images they created. They explore how popular and scientific representations of dinosaurs have influenced our changing perceptions of dinosaurs and the world in which they lived.

20

If you love those lawyer-eating, King Kong-battling movie dinosaurs, **Glut**'s article is sure to bring back happy memories of popcorn, soda, and Saturday matinees. He gives readers a thorough review of cinematic dinosaurs, from animated *Gertie* to the modern films of the Spielberg team and Disney Studios. In the process, he shows readers how scientific ideas of the day were incorporated in these films, as well as how common misconceptions about the "terrible reptiles" are often perpetuated on film.

Putting the Pieces Together: Dinosaur Restoration

17

Robert F. Walters and Tess Kissinger

Walters & Kissinger LLC, Philadelphia, PA

One of the most interesting professions a person can have is that of a paleontological life restoration artist. Usually referred to simply as "paleo artists," they work with scientists to try and solve the mystery of what dinosaurs may actually have looked like as living animals in their particular environments. A paleo artist usually has a college background in the techniques of drawing, painting, and sculpture, and has also intensively studied the sciences of comparative animal anatomy, dinosaur paleontology, biomechanics, and paleobotany. Although many artists accompany paleontologists on field trips and do drawings at dig sites, the real work of life restoration begins after the fossil evidence is returned to the museum or university and the bones are prepared for study.

The paleo artist begins the task of "bringing dinosaurs back to life" by examining the bones. The best way to do this is to visit the collection where the fossil bones have been carefully cleaned and separated from the rock in which they were found, but sometimes this is not possible. When the fossil evidence is, for whatever reason, unavailable, the paleo artist begins with all of the scientific information about the animal that is accessible. This information may include casts of the bones, as well as any scientific papers with pictures and measurements of the fossilized bones and other information about their discovery. The paleo artist also studies closely related species, interviews the paleontologist who discovered and described the dinosaur in question, and talks with other paleontologists specializing in dinosaurs similar to the one being portrayed.

Paleo artist Robert Walters completed a life restoration of *Giganotosaurus carolinii*, a recently discovered, large, carnivorous dinosaur from South America (Fig. 1). Bob worked closely with Dr. Rodolfo Coria, the paleontologist who first described *Giganotosaurus*, to get the measurements of the fossil bones correct. After studying Professor Coria's paper on *Giganotosaurus*, having several discussions with the professor, and comparing the skeleton with

> *Doing paleo art is like doing an autopsy in reverse—rather than dissect an animal to search for the cause of death, the artist seeks to "flesh out" the animal to recreate what it looked like in life.*
>
> *Walters & Kissinger*

Robert F. Walters knew he wanted to be a dinosaur artist when he was four years old, after seeing Zallinger's great mural, "The Age of Reptiles," on the cover of *Life* magazine. He began drawing dinosaurs immediately, studied art at the Academy of Fine Arts in Philadelphia, and has been a professional dinosaur life restoration artist for more than 20 years. His artwork is on permanent display in museums across the country, and he has illustrated more than 20 dinosaur books and innumerable magazines.

Tess Kissinger has been an artist and art administrator for more than 20 years. She is the author of the book *Copyrights, Contracts, Pricing and Ethical Guidelines for Dinosaur Artists & Paleontologists*, published by The Dinosaur

continued on p.151

Fig. 2. Skull of *Giganotosaurus* and probable muscle placement.

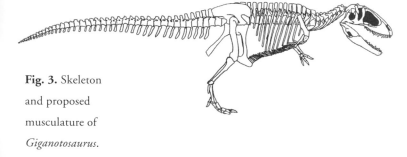

Fig. 3. Skeleton and proposed musculature of *Giganotosaurus*.

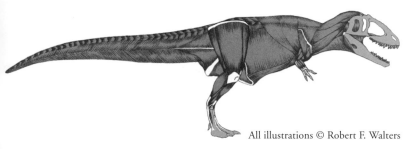

All illustrations © Robert F. Walters

other related carnivorous dinosaurs, Bob prepared an accurate skeletal reconstruction.

Looking at the bones of similar dinosaurs can be very important since it is extremely rare for complete skeletons to be recovered—the *Giganotosaurus* was only 70 percent complete. But Bob was lucky. With *Giganotosaurus*, often a missing bone on one side of the animal could be modeled from its existing counterpart on the other side. This practice is called "mirror-imaging." Paleo artists are not always this lucky. Bob also benefited from having a cast of the *Giganotosaurus* skull in our studio from which he could make detailed drawings (Fig. 2). These technical drawings are used in scientific papers, providing information about the particular animal to other scientists and artists (Fig. 3).

Doing paleo art is like doing an autopsy in reverse—rather than dissect an animal to search for the cause of death, the artist seeks to "flesh out" the animal to recreate what it looked like in life. With a good understanding of anatomy and the fossil bones to study, paleo artists can locate the sites of muscle attachment and then place muscles onto the skeleton.

Doing a life restoration also involves recreating the environment in which the animal lived. The artist consults with paleobotanists to find out what plant species would have been in the animal's surroundings (Fig. 4).

A life restoration of a large plant-eating dinosaur uses the same methodology. We had the pleasure of bringing *Gryposaurus notabilis*, a large duckbill dinosaur from the late Cretaceous of North America, "back to life" (Fig. 5). Once again, we were very lucky to work with a nearly complete fossil skeleton, though there are many other specimens of duckbill dinosaurs with which we were able to make comparisons. During the entire process of restoring

Fig. 4.
Life restoration
of two
*Giganotosaurus*es
in the Creta-
ceous of
Argentina, 90
million years
ago.

Gryposaurus, we frequently consulted with paleontologists to determine the animal's appearance, stance, possible behaviors, and surrounding environment (Fig. 6).

Regarding appearance, we know that other closely related species of dinosaurs probably had horny beaks, though the horny material of the beak is seldom preserved. Although no beak was found with *Gryposaurus*, we felt comfortable in portraying the dinosaur with a beak. As for stance, we know from comparing the lengths of the arms and legs of *Gryposaurus* and from fossil trackway evidence, that it had the ability to walk on two legs or on all four limbs. From fossil nesting sites from other hadrosaur species, such as *Maiasaura*, we suspect that there may have been parenting behavior in these dinosaurs. Finally, the fossil record indicates that the landscape of the late Cretaceous had many of the plant species we see today, so the environment

Fig. 5. Skull and proposed head musculature of *Gryposaurus notabilis.*

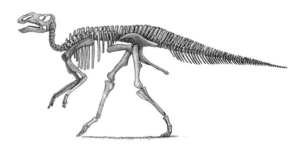

Fig. 6.
Three steps:
skeleton recon-
struction,
musculature,
and life
restoration of
*Gryposaurus
notabilis.*

Fig. 7. A *Gryposaurus* family in its environment.

Fig. 8. The mounted skeleton of *Tyrannosaurus rex* at the Academy of Natural Sciences in Philadelphia.

Fig. 9. Paul Sorton's life restoration model of *Tyrannosaurus rex*.

would look far more familiar (Fig. 7) than that surrounding *Giganotosaurus*.

Many paleo artists choose to sculpt, rather than paint their life restorations and their works can be anywhere from tiny to life-sized (Fig. 8). Paleo artist Paul Sorton took detailed measurements of the entire skeleton of *T. rex* by climbing on ladders and using a Genie lift. Those measurements were then scaled down to create a model *Tyrannosaurus* 1/24 the actual size (Fig. 9).

Fig. 10. Simple 3-D digital model of *T. rex*.

The computer is a powerful new tool, allowing paleo artists to utilize three-dimensional digital information to create models of dinosaurs. Computer animator Harry Saffren created a computer model (skinless) of *Tyrannosaurus rex* (Fig. 10), working from two-dimensional drawings like those in Figs. 3 and 6. Such models can then be "painted" (wrapped in a digital skin) and animated to move in any position through a three-dimensional landscape, just as they do in the movie *Jurassic Park* and the television special *Walking With Dinosaurs*. The computer also allows us to test certain kinds of animal movements and locomotion to see if they are biomechanically feasible.

How does the paleo artist know what dinosaur skin looked like? Paleontologists have found preserved skin impressions from a number of different species of dinosaurs. There are examples of skin

continued from p. 147

Society. Her work has been featured in exhibits at several museums, including the Bruce Museum and the Smithsonian Institution. She is a member of the American Society of Appraisers and is the world's foremost specialist in the appraisal of historical and contemporary dinosaur art.

Fig. 11. "Mummified" hand of *Edmontosaurus*.

Fig. 13. Fossil of *Sinosauropteryx* showing a covering of downy feathers.

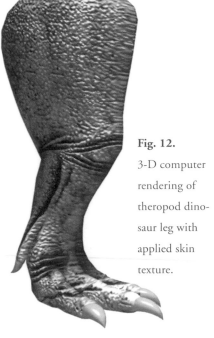

Fig. 12. 3-D computer rendering of theropod dinosaur leg with applied skin texture.

All illustrations © Robert F. Walters

impressions from sauropods, theropods, ornithopods, and ceratopsians. Skin impressions from hadrosaurian dinosaurs, like *Edmontosaurus* (Fig. 11), show rounded pebble-like scales covering large portions of the body, so we chose to use this type of skin texture in our reconstruction of the closely related dinosaur, *Gryposaurus*. Skin impressions from the theropod *Carnotaurus* provide paleo artists with a model for the scale patterns of other large carnivorous dinosaurs. Artists Harry Saffren and Bob Walters used the *Carnotaurus* model in rendering a skin texture for the leg of *Tyrannosaurus rex* (Fig. 12).

For more than two decades, it has been suspected that many small, carnivorous dinosaurs may have had feathers. Recent discoveries of early Cretaceous dinosaurs in China support this hypothesis (Fig. 13). Paleo artist Bob Walters, using photographs he took of the specimen, generated a life restoration of *Sinosauropteryx* (Fig. 14).

Fig. 14.

Life restoration of *Sinosauropteryx.*

Fig. 15.

Two different color patterns for *Triceratops.*

One of the questions most often asked of paleo artists is, "How do you know what colors dinosaurs were?" The simple answer is, we don't know. There is no preserved information about dinosaur coloration. Paleo artists can use modern reptiles and birds as guides because they are the dinosaurs' nearest living relatives. But this poses a problem. Many modern reptiles and birds are brightly colored, but most very large animals alive today have dull coloration. So, were dinosaurs dull or brightly colored (Fig. 15)? This much is left to the paleo artist's imagination, although it is a good bet that dinosaurs weren't plaid!

Dinosaurs dominated terrestrial life millions of years ago, and now paleo artists and paleontologists work together like detectives, sifting the fossil evidence to solve the mystery of what dinosaurs looked like. With each new paleontological discovery, information is added to the story of what life was once like. It is the paleo artists who are privileged to illustrate that story.

All illustrations © Robert F. Walters

Dinosaur Superlatives

Peter Dodson

School of Veterinary Medicine
University of Pennsylvania, Philadelphia

18

Dinosaurs are among the most successful animals that have ever lived. For 160 million years or more, they were the largest animals on land. They inhabited the landmasses that today constitute all seven continents. Part of the appeal of dinosaurs undoubtedly derives from their large size, yet the majority of dinosaurs were not all that big. On average, most dinosaurs were between the sizes of cows and elephants, certainly not 10 times the size of elephants. Some dinosaurs were not much bigger than chickens! Much about dinosaurs is true and beautiful apart from large size. Bigger is not necessarily better. Good things often come in medium size or even small packages, and the same is true of dinosaurs. Numerous superlatives are applicable to Dinosauria. We can begin at the beginning with some "dinosaur firsts."

> *. . . the majority of dinosaurs were not all that big. On average, most dinosaurs were between the sizes of cows and elephants . . .*
>
> Peter Dodson

Dinosaur Firsts

Which Was the First Dinosaur to Receive a Formal Description and Name?

Reverend William Buckland of Oxford named the Jurassic meat-eater *Megalosaurus* in 1824, based on fossils he found in 1818. He did not give it a species name. This oversight was corrected by a German scientist, Christian Erich Hermann von Meyer, in 1832; he honored the Reverend Buckland by naming the dinosaur *Megalosaurus bucklandi*. The second dinosaur to receive a description and generic name was the ornithopod *Iguanodon*, named in 1825 by Sussex physician Gideon Mantell. Mantell and his wife found the first *Iguanodon* fossils three years earlier, in 1822. He, too, failed to coin a species name for his dinosaur. In 1829, Holl named it *Iguanodon anglicus*, so the second dinosaur to be named was actually the first to receive a complete designation.

When Were Dinosaur Bones First Observed?

This question is hard to answer. It is most likely that bones were observed very early on, yet not identified, nor recorded. Classicist and folklorist Adrienne Mayor believes that Scythian gold miners, crossing the Gobi Desert

Peter Dodson received his Ph.D. in Earth Sciences from Yale University in 1974. He has spent his entire career as a gross anatomist at the University of Pennsylvania School of Veterinary Medicine, with a secondary appointment in the Department of Geology. He has done extensive fieldwork in the western United States and Canada, with additional fieldwork in China, India, Madagascar, Egypt, and Argentina. His books include *The Horned Dinosaurs* (1996) and *An Alphabet of Dinosaurs* (1995). He taught a Templeton course on science and religion at the University of Pennsylvania in 1999, and is past president of the Philadelphia Center for Religion and Science.

towards the Altai Mountains in Roman times, may have been the first to observe bones of what we now know to be dinosaurs.[1] The white bones would have stood out in stark contrast against the red sediments of the Djadoktha Formation, and these sightings may have given rise to the legend of the griffin, a mythical beast with the beak of an eagle and the claws of a lion—a description not completely unlike the dinosaur, *Protoceratops*, whose bones are common in this area. It is equally possible that Native Americans and other aboriginals observed dinosaur bones in the ground, recognized that they were the remains of very large extinct animals, and wove them into their legends and myths.

What we can say for sure is that the first verified report of dinosaur bones came from Western Europe and was recorded by Robert Plot in 1677 in the *Natural History of Oxfordshire*. This monograph contains an illustration of a bone that is unmistakably the lower end of the femur of a large meat-eating dinosaur, perhaps *Megalosaurus*. He thought it must have been "the Bone of some Elephant, brought hither during the Government of the Romans in Britain."[2] In 1763 Richard Brookes reillustrated the same specimen under the caption *Scrotum humanum*, in reference to its perceived resemblance to a particular part of the male anatomy. Another early dinosaur discovery came from Normandy, France. In 1801, the great French anatomist, Georges Cuvier, described this fossil as a crocodilian. We now recognize that he had described the braincase of a Jurassic theropod dinosaur.

When Did the Name Dinosaur First Appear?

In 1842. The name, according to its author, the great British anatomist, zoologist and paleontologist, Richard Owen, means "fearfully great reptile."

When and Where Did the Public First Learn About Dinosaurs?

In 1854 a series of life-sized dinosaur sculptures was unveiled at the Crystal Palace at Sydenham, just outside London. These models were the result of collaboration between Richard Owen, the scientist, and Benjamin Waterhouse Hawkins, the artist. The effort was a serious one. The trouble was that, up to this point, all dinosaur finds were based on fragmentary and extremely incomplete remains. The sculptures incorporated the concepts "large" and "reptilian," but most of the details were imaginary. The statues are bizarre to our eyes, but they had an electrifying effect on the public, and dinosaurs have fired the public imagination ever since.

When Were Dinosaurs First Discovered in the United States?

One may guess that Native Americans were the first to observe dinosaur bones, but any such finds have been obscured by the mists of time. The first dinosaur bone to receive scientific attention in the United States was probably a hadrosaur bone from Woodbridge, New Jersey, described by Caspar Wistar at a meeting of the American Philosophical Society in Philadelphia in 1787. Of course, it was not identified as a dinosaur at that time, because dinosaurs were not yet recognized. Dinosaur footprints in the Connecticut Valley of western Massachusetts attracted the attention of Pliny Moody as early as 1802. The first American fossils recognized as dinosaurs are four dinosaurs discovered in 1856 from what is now the Judith River Formation of Montana. These fossils were sent by Ferdinand Hayden back to Philadelphia, where Joseph Leidy described them as *Trachodon*, *Troödon*, *Deinodon* and *Paleoscincus*. As these fossils were only teeth, we cannot attribute these animals to any particular genera represented by skulls or skeletons, and except for *Troödon*, the names are no longer in use.

What Was the First Dinosaur Skeleton to Be Exhibited?

Hadrosaurus, the first of the hadrosaurs or duckbilled dinosaurs known to science, was described by Joseph Leidy in 1858. Ten years later it was exhibited as a freestanding skeletal mount (with liberal reconstruction of missing parts) at the Academy of Natural Sciences in Philadelphia. It was not until the first years of the 20th

century that other museums in the United States began mounting skeletons of dinosaurs.

When Were Dinosaurs First Found in Other Countries Around the World?

The first dinosaur discovered in Canada was excavated in 1884 near Drumheller, Alberta, by Joseph Burr Tyrrell of the Geological Survey of Canada. In 1905, Henry Fairfield Osborn named it *Albertosaurus sarcophagus*. The first dinosaur from Africa was the prosauropod *Massospondylus* from South Africa, named by Richard Owen in 1854. Argentina's first dinosaur was the sauropod *Argyrosaurus*, described by Richard Lydekker in 1893. Mongolia's first dinosaur was *Protoceratops*, described by Granger and Gregory in 1923. China's first dinosaur was the sauropod *Euhelopus*. It was first described as *Helopus* by the Swedish paleontologist Carl Wiman in 1929, but the name was found to already belong to another animal, so it was renamed in 1956 by A.S. Romer. The first dinosaur from Australia is the poorly known Middle Jurassic sauropod *Rhoetosaurus*, described in 1925 by Longman. The first dinosaur to be named from Antarctica is *Cryolophosaurus* (whose name most graphically means "cold crested reptile"). It is an Early Jurassic meat-eater, named in 1993 by W.R. Hammer and W.J. Hickerson.

Dinosaur Geography

Dinosaurs have been found all over the world. What superlative awards can we make in the geography category? Let's take a look.

Which Country Has the Most Genera of Dinosaurs?

The United States takes the gold in this category, and by a rather wide margin. A census based on data available in 1988 showed that six countries account for about 75 percent of all dinosaurs in general.[3, 4] In order, they are

United States	64 genera
Mongolia	40
China	36
Canada	31
England	26
Argentina	23

Just seven years later, a fresh census[5] showed that the United States had gained fifteen genera, Argentina eight, China eight, and Mongolia six. Impressive, but such data are pretty much out of date as soon as they are published, as new dinosaurs have been described at the rate of six or more per year since 1970. The flood of new dinosaurs shows not the slightest sign of abating in the United States or anywhere else.

Why Does the United States Have so Many Genera of Dinosaurs Compared to Other Countries?

Two factors are crucial here: (1) the United States has a very large surface area of arid and semi-arid lands where erosion predominates and new fossils are rapidly exposed. In England, humid conditions and luxuriant plant growth soon cover any natural exposures. (2) Perhaps most importantly, the United States has a lot of exposed rocks that are the right age—Late Triassic to Late Cretaceous, the time of the dinosaurs. In Canada and Mongolia, dinosaur-bearing rocks are almost exclusively Cretaceous in age, indeed principally Late Cretaceous—only a tiny slice of "dinosaur time." Two countries that may one day surpass the United States in the number of genera of dinosaurs are China and Argentina—both share with the United States a wide stratigraphic range of terrestrial fossil-bearing rocks of the appropriate geologic age.

Other factors may also play a role. These include (a) ecological conditions during the Mesozoic Era were highly favorable for both dinosaur diversity and eventual preservation across wide areas of the North American continent; (b) dinosaurs have been studied in the United States since 1856, a significantly longer period of time than in all the other countries listed with the exception of England; and (c) the United States has abundant financial resources and a large population of researchers.

Sauropod Record Holders

Before we examine the record holders, let us briefly review the major groups of dinosaurs. The first split is into the Saurischia or "lizard-

hipped" dinosaurs and the Ornithischia or "bird-hipped" dinosaurs. Saurischians include (1) the long-necked plant-eaters, the sauropods, (2) their precursors, the prosauropods, and (3) the meat-eating, two-legged theropods. Birds are direct descendants of small theropods. All ornithischians are plant-eaters. The four-legged variety includes (1) armored ankylosaurs, (2) plated stegosaurs, and (3) the horned ceratopsians, whose precursors were actually two-legged. A fourth group of ornithischians, the ornithopods, are two- or four-legged herbivores of small to large size and include the hadrosaurs or duck-billed dinosaurs.

Who's the Biggest of Them All?

The largest of all dinosaurs are the sauropods, the familiar long-necked, long-tailed herbivores, sometimes called "brontosaurs"—a term used in a well-intentioned, if technically incorrect, attempt to retain the beloved name of *Brontosaurus*. This particular dinosaur is known not only for its size, but also for its name change. Unfortunately, it was named twice and, interestingly enough, by the same person. In 1877, Yale University paleontologist O.C. Marsh first selected the name *Apatosaurus* for a new fossil find. Ironically, *Apatosaurus* means "deceptive reptile." Two years later, he applied the name *Brontosaurus*, meaning "thunder lizard," to a much larger, more complete and more spectacular—but headless—specimen, thinking it was a different dinosaur genus altogether. Eventually, additional fossil finds included skull elements, revealing that these two dinosaurs were, in fact, one in the same. The rules of zoological nomenclature are very clear on the point: each animal may have only one name, and the first name to be applied sticks. Thus *Brontosaurus* is now forever known as *Apatosaurus*.

Not all sauropods were giants, but all truly giant dinosaurs were sauropods. For many people, the essence of "dinosaur" is a sauropod. Truth be told, sauropods were somewhat archaic dinosaurs, Model Ts that lived in a world of turbochargers. Sauropods had small heads (the largest were roughly the size of horse heads), simple teeth, and the smallest brain-to-body-size ratio of all dinosaurs. Sauropods rose

to numerical dominance in the Jurassic, but by the latter half of the Cretaceous the fossil record reveals that they had been largely replaced on northern continents, especially North America, by more advanced herbivores. These were the ornithischians with larger heads, more complex teeth, and larger brains. Sauropods continued to dominate southern continents throughout the Cretaceous, and as a group were extremely successful, whatever their shortcomings.

So, which was the largest dinosaur? Nobody knows. At this point, all that we know is that the largest dinosaur was a sauropod.

So What Do We Mean by Large?

There are different components of "large." Do we mean the longest? The heaviest? The tallest? Size is an important factor. Should we insist that the dinosaur fossil be complete before it is allowed to enter the contest, or will we accept an estimate based on a single bone?

A sad fact of paleontological life is that most dinosaur fossils are woefully incomplete, making it a challenge sometimes to measure size, regardless of how we use the term.

Burying the carcass of an animal the size of a sauropod obviously takes time. Before burial is accomplished, there is often ample time to lose portions of the tail, feet, head, and neck to the processes of physical, chemical, and biological disintegration. Only a few sauropods are known in their entirety (e.g., *Camarasaurus, Diplodocus, Shunosaurus, Mamenchisaurus*). Several other sauropods are well known, but with a few details missing (*Apatosaurus, Brachiosaurus*). The complete or almost complete sauropods serve as templates for estimating dimensions of sauropods that are less complete, or known only from a few bones. Looking, then, at only essentially complete specimens: (a) the longest dinosaur skeleton belongs to *Diplodocus*, at 27.4 m (90 ft), and (b) the tallest is that of *Brachiosaurus brancai*. The skeleton of this animal, as mounted at the Humboldt Museum in Berlin, is 11.9 m (39 ft) high, though it is only 22.5m (74 ft) long.

What we do know, however, is that these are, in fact, probably not the winning numbers. For instance, Gregory Paul points out that other bones of *Brachiosaurus* in the Humboldt

collection are roughly 10 percent larger than those of the mounted specimen.[6] That could indicate that *Brachiosaurus* was as much as 13.1 m (43 ft) high with the neck elevated, and 24.7m. (81 ft) long. In fact, the title of largest of the large may well belong to an entirely different dinosaur.

Serious claims have been advanced for *Seismosaurus* as the world's longest dinosaur (Fig. 1). *Seismosaurus* is a diplodocid from the Morrison Formation of New Mexico.[7] This partial articulated skeleton includes dorsal vertebrae, sacrum, pelvis, and base of the tail out to the 27th caudal vertebra. Based on comparisons of the linear measurements of preserved skeletal elements with those of *Diplodocus*, the length of *Seismosaurus* has been estimated at between 39 and 52 m (128 ft and 171 ft). No competing claim has been made for greater length—thus far.

How Do Dinosaurs Weigh In?

Estimation of body weight can be difficult. Several methods have been used, with varying success. One method involves the building of accurate scale models, then measuring the volume of the model by displacement of water or sand.[8, 9] From this figure, one calculates the volume of the living animal by scaling, making an assumption about the specific mass of the living animal. Thus, one arrives at body weight. We do need to be aware that models differ according to how much bulk the model-maker restores—some artists like their dinosaurs thin and gaunt, others prefer roly-poly.

Another method, developed by Anderson and colleagues for living mammals, uses equations based on the cross-sectional area of the humerus and femur. This method calculates weight based on the relationship of these variables to body mass.[10] In the case of either method, body weights calculated for dinosaurs are estimates, not facts.

Based upon estimation, *Argentinosaurus* may well be the heavy weight of the dino-

saurs.[11] Preserved materials of this immense titanosaur from the middle Cretaceous of Argentina include six dorsal vertebrae, some ribs, and a partial sacrum. There is also a fibula measuring 155 cm in length, which would cor-

© 1999 Patti Kane-Vanni

respond to a femur length of approximately 2.4 m, although no such enormous femur has yet been found from here or anywhere else. Weight estimates for *Argentinosaurus* range from 73,000 kg (80 tons)[12] to 90,000 kg (100 tons).[8]

However, there may be another contender for the title. The heaviest of all sauropods may be *Amphicoelias fragillimus*, at 125,000 kg (137.5 tons). This sauropod, from the Late Jurassic Morrison Formation of Colorado, is among the most poorly known of the poorly known sauropods. The only specimen we had was the neural arch of a dorsal vertebra, which has been lost. E.D. Cope, who described the specimen in 1878, believed that the height of the intact vertebra would have exceeded 1.8 m (6 ft). It is difficult to say much about such an insubstantial taxon.

What About the Bits and Pieces?

If we look only at the isolated elements that have been found thus far, we can begin an entire new list of dinosaur superlatives from head to tail (Table 1).

What About the Little Guys?

Sauropods are justifiably famous for their enormous sizes, but even *Seismosaurus* was once

Fig. 1.

A 135-foot skeletal reconstruction of *Seismosaurus* on the grounds of The Wyoming Dinosaur Center, Thermopolis, Wyoming. Pictured are members of Dodson's 1999 Montana field team for scale.

Table 1. A sampling of some of the longest and largest isolated bone elements

Longest neck	9.8 m (33 ft)	*Mamenchisaurus hochuanensis* from the Late Jurassic of Sichuan Province, China
Largest shoulder girdle	2.69 m (8 ft 10 in)	*Brachiosaurus*, from Dry Mesa in western Colorado, collected by Jensen in 1985
Longest cervical vertebra	1250 mm (49 in)	*Sauroposeidon*, from Oklahoma, described in 2000 by Wedel and colleagues[13]
Longest cervical rib	4.1 m (13 ft 5 in)	*Mamenchisaurus sinocanadorum*, from China, described by Russell and Zheng, 1993[14]
Longest humerus	2.04 m (80 in)	*Brachiosaurus altithorax*, from western Colorado, collected by Elmer Riggs for the Columbian field Museum in 1900 (for the second place finisher, see Fig. 2)
Longest femur	2.3 m (7 ft 6 in)	*Antarctosaurus giganteus*, from the early Late Cretaceous of Argentina, described by Huene in 1929 (Fig. 3)

small enough to fit in an egg. Baby sauropods hatched from eggs that were spherical or subspherical in shape. Eggs attributed to sauropods usually do not exceed two liters in volume, about half the maximum size of dinosaur eggs so far reported. Sauropod eggs are common at a number of sites around the world, but the discovery of a titanosaur nesting site by Chiappe and others in Argentina is particularly important.[15] The eggs at this site are 13 to 15 cm (5 to 6 in) in diameter and contain the remains of embryonic sauropods. What makes these eggs even more spectacular is that the embryos show patches of scaly skin. Association of sauropod bones with eggs elsewhere in the world is unproven.

We have fossils of baby sauropod specimens from several places around the world. This means that we are examining some really small fossils—some may even be from embryonic dinosaurs or very young hatchlings. So for smallest dinosaurs, we have to consider the following two specimens: (1) A partial jawbone of tiny *Camarasaurus* that calculations suggest would correspond to a body length of only 1.1 m (43 in);[16, 17] and (2) *Mussaurus*, the "mouse reptile," a Late Triassic prosauropod from Argentina described by Bonaparte and Vince in 1979. The skull of this hatchling (or possibly embryo) is only 32 mm (1.25 in) long, and the

femur 30 mm (1.15 in). Therefore, the overall length of the baby is estimated to be only 25 or 30 cm (10 to 12 in).

Maybe describing baby dinosaurs under the "smallest" superlative is a bit unfair. What about the smallest adult sauropod? This honor may go to *Magyarosaurus dacus* from Romania—a mere 5 or 6 m (17–20 ft) long.[18]

Theropod Record Holders
Which Was the Largest Meat-Eater?

For many years everyone knew that *Tyrannosaurus rex* was the biggest, baddest predator that ever lived. Its skull was 1.4 m (4.5 ft) long and its teeth were up to 15 cm (6 in) high. *T. rex* was discovered in Montana in 1902 by Barnum Brown, and described in 1905 by Henry Fairfield Osborn. Its skeleton dominated the Cretaceous Hall at the American Museum of Natural History, standing 4.6 m (15 ft) high and stretching out 15 m (50 ft).

However, in recent years, this admitted monster has had a bit of come down.[19] Additional fossil evidence has resulted in adjustments to skeletal mounts. The maximum skull length has increased by nearly a foot, but we've chopped its tail by about 12 feet. To add insult to injury, several potential rivals have hit the scene.

1) In 1915 Ernst Stromer collected the poorly preserved remains of a very large sail-backed predator, *Spinosaurus*, from Late Cretaceous rocks in Egypt. We have neither a complete skull nor any long bones, so we cannot specify its exact size, but based on the bones that were recovered, *Spinosaurus* was at least close to *Tyrannosaurus* in size, and may well have been longer.

2) A second possible rival, found in Egypt and in Morocco, is the very large theropod named *Carcharodontosaurus*, the "shark-toothed reptile."[20] Its skull has been reconstructed by Paul Sereno at about 1.8 m (6 ft) long, clearly larger than the largest known specimens of *Tyrannosaurus*.

3) In 1934, Stromer described a third contender, *Bahariasaurus*, also from Egypt. This animal too is a *Tyrannosaurus*-sized predator.

4) An additional giant predator, *Giganotosaurus*, was described from the middle Cretaceous of Argentina, by Coria and Salgado.[21] As reconstructed, *Giganotosaurus* is 12.5 m (42 ft) long, and has a skull 1.8 m (6 ft) long. The case may be made that it, not *Tyrannosaurus*, is the largest predator.

Regrettably, all of the three Egyptian specimens were destroyed during an Allied bombing run over Munich in 1944. However, based upon both the literature and the remaining fossil materials, we can say that the top predators on three different continental masses (Africa, North America, and South America), all appeared to be about the same size—big!

What About the Little Theropods?

Inasmuch as birds are recognized to be theropods, the smallest theropod is a hummingbird! However, if consideration is limited to non-avian theropods, the choice becomes a little clearer. The smallest known theropod comes from the Early Cretaceous of northeastern China, the site of so many spectacular finds of primitive birds and near-birds. The title-holder thus far is *Microraptor*, described in 2000 by Xu Xing and colleagues.[22] This dinosaur appears to be a tiny adult member of the

Dromaeosauridae. Its trunk length is stated to be 47 mm (1.9 in), thus much smaller than any known specimen of the earliest bird, *Archaeopteryx*.

Fig. 2. The 1.69 m (5 ft 7 in) humerus (the second-largest yet discovered), from *Paralititan stromeri*, excavated near the Bahariya Oasis in Egypt's Western Desert in 2000. Pictured are members of the field team from "The Bahariya Dinosaur Project": (L to R) Jen Smith, Jason Poole, Patti Kane-Vanni, Matt Lamanna, Josh Smith and Dr. Peter Dodson.

Ornithischian Record Holders

Size-wise, this group of dinosaurs does not create headlines. However, it is a group that can claim a few prizes in at least two special categories: spectacular eggs and head size.

Who Laid the Biggest Egg?

Some of the most spectacular dinosaur fossils found in North America are those of baby hadrosaurs and their eggs. The eggs of the hadrosaur *Hypacrosaurus stebingeri* from extreme southern Alberta and adjacent Montana are among the largest dinosaur eggs known.[23] Each egg had a volume of nearly four liters, with a diameter measuring about 20 cm (8 in). Size of the hatchlings was probably somewhat less than 1 m; a rule of thumb is that length of the hatchling is roughly three times maximum diameter of the egg. The spherical eggs of another hadrosaur from Montana, *Maiasaura*, are much smaller, according to Horner, with an es-

timated volume of 900 cc and a diameter of 12 cm (5 in).[24] Hatchlings would have been 33 to 45 cm (13 to 18 in) in length and would have weighed perhaps 800 g (1.75 lbs).

Fig. 3.

Six-foot one-inch student Matt Lamanna next to the 2.3 m femur of the *Antarctosaurus giganteus* in the Museo La Plata, Argentina.

© 1999 Peter Dodson

What About Head Size?

It has long been appreciated that the skulls of the large ceratopsids are not only the largest skulls of any dinosaurs, they are also the largest skulls of any land animals. Some truly enormous skulls have been found. Lehman described a *Pentaceratops* skull from the Late Cretaceous of New Mexico that was estimated at 3 m (9 ft 10 in) in length.[25] This specimen is, so far, without rival in size. Oddly, the skeleton accompanying the skull is not of exceptional size, so the overall length of the animal was 6.8 m (22 ft 4 in), well off the record size for *Triceratops* at 7.9 m (26 ft).

The opposite end of the size spectrum is still up for grabs. It may belong to *Graciliceratops* ("slender horned face") from Mongolia.[26]

This animal is about 90 cm (35 in) long, 26 cm (10 in) high at the hips, and had a skull 20 cm (8 in) long. If the specimen is adult, it may have been the smallest dinosaurian herbivore. However, one of the smallest dinosaur skulls in existence is only 28 mm (1.1 in) long. It is the skull of *Psittacosaurus*, most likely an embryonic or newly hatched individual.

Head of the Class

Early portrayals described dinosaurs as slow, lumbering, dim-witted beasts, regardless of skull size. In contrast, the *Velociraptor* in *Jurassic Park* nimbly darted about using keen senses to track down its terrified prey. So how "brainy" were these superlative creatures, and how can we tell?

Who Had the Biggest Brain?

Without doubt the "biggest and brightest" prize would not go to the sauropods! They hold the unenviable superlative title of smallest brain, relative to body size, of any of the dinosaurs, with only about 20 percent of the brain volume that would be expected of an alligator of sauropod size![27, 28]

Brain volume is determined by measuring the actual size of the brain case within the skull; new technologies enable us to do this quite accurately (see Chapman, page 137). If we look at brain volume alone, then *Tyrannosaurus* is the winner. It has the largest reported brain cavity volume of any dinosaur, approaching half a liter, depending on precisely what is measured. Assuming that the brain occupied the entire cranial space, its brain size would have exceeded that of a chimpanzee, a gorilla, or even that of a fossil hominid such as *Australopithecus*. However, *Tyrannosaurus* weighed 6,000 kg or more. We would expect a large skull and a large brain.

Does Big Mean Brainy?

In order to determine which dinosaur is the brainiest, we have to look at its EQ—not its IQ, its EQ—or encephalization quotient. The EQ is determined by comparing a ratio of brain volume/body size, to the same ratio in an

alligator that has been scaled to equal size.[27, 28] Using this method, if we look back at our prize-winning *Tyrannosaurus* and conservatively calculate its brain volume at 50 percent of brain cavity volume, then this dinosaur earns an encephalization quotient (EQ) of 1.04, meaning that its brain was 1.04 times the size of the brain of an alligator of *Tyrannosaurus* size.

The smaller *Allosaurus* proves to be brainier, having an EQ of 1.90, almost twice that of an alligator.

But without doubt, small theropods are the brainiest, which is to say, the most encephalized, of the dinosaurs. The underside of the frontal and parietal bones roofing the braincases of some small theropods show round concavities for swollen cerebral hemispheres. These concavities increase the brain volume. *Troödon*, in particular, has an extraordinary EQ of 5.8, which is comparable to that of living ostriches and opossums (but emphatically not that of chimpanzees). Brain volumes and EQs have not yet been reported for any other maniraptoran, but increased brain size is likely a general character of small theropods. However, precise measurements and detailed studies lie in the future.

Some interesting things happen when we look more closely at the plant-eating dinosaurs. Brain sizes of ceratopsians appear to have been modest. The best that can be said is that they had the largest EQs of any of the fully quadrupedal plant-eaters, though the published determinations (.88 for *Protoceratops*; .67 for *Triceratops*) are certainly less than the alligator index of 1.00. However, for those dinosaurs that probably spent at least some time on two legs (facultative bipeds), we see something quite different. Hopson reports the EQ of *Iguanodon* as 1.50 and of *Edmontosaurus* as 1.30, both beating out *Tyrannosaurus!*[27, 28]

Looking at Superlatives

So many categories, so little space! Who had the biggest teeth? Who sported the most unusual arrangement of head adornments? Who lived the longest? Who traveled the furthest? Depending upon the question asked and who is doing the judging, the competition becomes as fierce as some of the dinosaurs themselves. We all have our favorites, but what remains obvious is that new discoveries will be made, and not a single dinosaur is or was "the best."

R e f e r e n c e s

1. Mayor, A. 2000. *The first fossil hunters—paleontology in Greek and Roman times*. Princeton: Princeton University Press.

2. Sarjeant, W.A.S. 1997. The earliest discoveries. In *The complete dinosaur*, eds. J.O. Farlow and M.K. Brett-Surman, 3–11. Bloomington: Indiana University Press.

3. Dodson, P. 1990. Counting dinosaurs: how many kinds were there? *Proceedings of the National Academy of Sciences* 87:7608–7612.

4. Dodson, P., and S.D. Dawson. 1991. Making the fossil record of dinosaurs. *Modern Geology* 16:3–15.

5. Holmes, T., and P. Dodson. 1997. Counting more dinosaurs: how many kinds are there (1996)?. In *Dinofest international, proceedings of a symposium*, eds. D.L. Wolberg, E. Stump, and G.D. Rosenberg, 125–128. Philadelphia: Academy of Natural Sciences.

6. Paul, G.S. 1988. The brachiosaur giants of the Morrison and Tendaguru with a description of a new subgenus, *Giraffatitan*, and a comparison of the world's largest dinosaurs. *Hunteria* 2:1–14.

7. Gillette, D.D. 1991. *Seismosaurus halli*, gen. et sp. nov., a new sauropod from the Morrison Formation (Upper Jurassic/Lower Cretaceous) of New Mexico, USA. *Journal of Vertebrate Paleontology* 11:417–433.

8. Paul, G.S. 1997. Dinosaur models: the good, the bad, and using them to estimate the mass of dinosaurs. In *Dinofest international, proceedings of a symposium*, eds. D.L. Wolberg, E. Stump, and G.D. Rosenberg, 129–154. Philadelphia: Academy of Natural Sciences.

9. Alexander, R.M. 1998. All-time giants: the largest animals and their problems. *Palaeontology* 41:1231–1245.

10. Anderson, J.F., A. Hall-Martin, and D.A. Russell. 1985. Long-bone circumference and weight in mammals, birds and dinosaurs. *Journal of Zoology, London A* 207:53–61.

11. Bonaparte, J.F., and R.A. Coria. 1993. Un nuevo y gigantesco sauropodo titanosaurio de la Formacion Rio Limay (Albiano-Cenomaniano) de la Provincia del Neuquen, Argentina. *Ameghiniana* 30:271–282.

12. Burness, G.P., J. Diamond, and T. Flannery. 2001. Dinosaurs, dragons, and dwarfs: the evolution of maximal body size. *Proceedings of the National Academy of Sciences* 98:14518–14523.

13. Wedel, M.J., R.L. Cifelli, and R.K. Sanders. 2000. Osteology, paleobiology, and relationships of the sauropod dinosaur *Sauroposeidon*. *Acta Palaeontologica Polonica* 45:343–388.

14. Russell, D.A., and Z. Zheng. 1993. A large mamenchisaurid from the Junggar Basin, Xinjiang, People's Republic of China. *Canadian Journal of Earth Science* 30:2082–2095.

15. Chiappe, L.M., R.A. Coria, L. Dingus, F. Jackson, A. Chinsamy, and M. Fox. 1998. Sauropod dinosaur embryos from the Late Cretaceous of Patagonia. *Nature* 396:258–261.

16. Britt, B.B., and B.G. Naylor. 2000. An embryonic *Camarasaurus* (Dinosauria, Sauropoda) from the Upper Jurassic Morrison Formation (Dry Mesa Quarry, Colorado). In *Dinosaur eggs and babies*, eds. K. Carpenter, K.F. Hirsch, and J.R. Horner, 256–264. New York: Cambridge University Press.

17. Carpenter, K., and J. McIntosh. 1994. Upper Jurassic sauropod babies from the Morrison Fm. In *Dinosaur eggs and babies*, eds. K. Carpenter, K.F. Hirsch, and J.R. Horner, 265–278. New York: Cambridge University Press.

18. Jianu, C.-M., and D.B. Weishampel. 1999. The smallest of the largest: a new look at possible dwarfing in sauropod dinosaurs. *Geologie en Mijnbouw* 78:335–343.

19. Dodson, P. 1997. *Tyrannosaurus rex*—plus ultra? *American Paleontologist* 5(4):5–7.

20. Sereno, P.C., D.L. Dutheil, M. Iarochene, H.C.E. Larsson, G.H. Lyon, P.M. Magwene, C.A. Sidor, D.J. Varricchio, and J.A. Wilson. 1996. Predatory dinosaurs from the Sahara and Late Cretaceous faunal differentiation. *Science* 272:986–990.

21. Coria, R.A., and L. Salgado. 1995. A new giant carnivorous dinosaur from the Cretaceous of Patagonia. *Nature* 377:224–226.

22. Xu, X., Z. Zhou, and X. Wang. 2000. The smallest known non-avian theropod dinosaur. *Nature* 408:705–708.

23. Horner, J.R., and P.J. Currie. 1994. Embryonic and neonatal morphology and ontogeny of a new species of *Hypacrosaurus* (Ornithischia, Lambeosauridae) from Montana and Alberta. In *Dinosaur eggs and babies*, eds. K. Carpenter, K.F. Hirsch, and J.R. Horner, 312–336. New York: Cambridge University Press.

24. Horner, J.R. 1999. Egg clutches and embryos of two hadrosaurian dinosaurs. *Journal of Vertebrate Paleontology* 19:607–611.

25. Lehman, T.M. 1998. A gigantic skull and skeleton of the horned dinosaur *Pentaceratops sternbergi* from New Mexico. *Journal of Paleontology* 72:894–906.

26. Sereno, P.C. 2000. The fossil record, systematics, and evolution of pachycephalosaurs and ceratopsians from Asia. In *The age of dinosaurs in Russia and Mongolia*, eds. M.J. Benton, M.A. Shishkin, D.M. Unwin, and E.N. Kurochkin, 480–516. Cambridge: Cambridge University Press.

27. Hopson, J.A. 1977. Relative brain size and behavior in archosaurian reptiles. *Annual Review of Ecological Systematics* 8:429–448.

28. Hopson, J.A. 1980. Relative brain size in dinosaurs: implications for dinosaurian ectothermy. In *A cold look at the warm-blooded dinosaurs*, eds. R.D.K. Thomas and E.C. Olson, 287–310. American Association for the Advancement of Science Selected Symposium 28.

The Art of Illustrating Science

19

Dale A. Springer
Department of Geography and Geosciences
Bloomsburg University, Bloomsburg, PA
Brian Platt
University of Kansas, Lawrence

Close your eyes. Now think about dinosaurs. Do words like diapsid, sigmoidal third metatarsal, and parasagittal posture come to mind? Probably not! You probably see a picture of an animal. It's probably big. It may have wickedly sharp, curved teeth and claws, or a long whip-like tail, or horns on its face. Maybe your dinosaur is just standing there, looking like some stolid Mesozoic statue (Fig. 1). Or maybe you see the animal in motion, leaping to attack or to defend itself, rising up on hind legs to reach vegetation high in a tree, or reaching down to feed its young (Fig. 2). If your vision is particularly detailed, you may see scaly skin and color patterns, or hear the animal roar. Why does the word "dinosaur" generate a particular picture in your mind? If we had asked someone in1854 to consider dinosaurs, they would certainly have described a very different creature from the ones you imagine. So, where have your perceptions of them come from? And, are they accurate?

Our perceptions are strongly shaped by images we see. Images are powerful—they stick with us. For most of us, our first memorable impressions of the world around us come through vision. This is as true for our perceptions of dinosaurs and the long gone landscapes they inhabited as it is for living organisms. Our popular view of dinosaurs has been shaped by the visual arts.

Dinosaur images abound in the media today. Think *Jurassic Park*, *Walking with Dinosaurs*, *Calvin and Hobbes*, or—if you have a very long memory—*The Lost World*. These images fire our imagination about a world more than 65 million years in Earth's past. Yet, it's a long way from scattered bones preserved in the mudstone of an ancient lakeshore to the images of the "fearfully great lizards" we see in magazines, museums, and Spielberg productions.

How do the artists who create dinosaur images decide what these extinct animals looked like? Have popular illustrations of dinosaurs changed since these animals first came to public attention 160 or so years ago? Are the im-

> *... it's a long way from scattered bones preserved in the mudstone of an ancient lakeshore to the images ... we see in magazines, museums, and Spielberg productions.*
>
> *Springer & Platt*

Dale A. Springer is a paleontologist and professor of Geosciences at Bloomsburg University in Pennsylvania. Her major research interests include the organization of marine invertebrate communities and investigating the processes responsible for changes in the spatial and temporal distributions of these communities. Dr. Springer has been involved in geoscience education initiatives for almost 25 years. She has served on several committees of the American Geological Institute (AGI), and is the current Education and Outreach Coordinator of the Paleontological Society (PS). She is an editor and co-author of a Paleontological Society publication, *Evolution: Investigating the Evidence*, and co-author of the booklet, *Evolution and the Fossil Record*, produced by the AGI and PS. She is also

continued on p. 167

ages we see today more accurate than those of 1854? Few of us spend much time thinking about these questions. More likely, we simply assume that pictures in books and magazines represent dinosaurs and their world as they

Fig. 1.

Gertie the Dinosaur, the first animated dinosaur, drawn by Winsor McCay for a short film (1914) of the same title. McCay drew Gertie with a remarkably upright stance and, in the film, she is shown behaving more like a terrestrial mammal than a semi-aquatic reptile, the common portrait of sauropods through much of the 20th century. Platt, after Winsor McCay.

truly were. But, is this a correct assumption? The question is not a trivial one. It goes to the heart of one of the most important goals of good science: conveying scientific discoveries and concepts to the general public.

Non-scientists rarely read scientific journals. Even scientists may not read journals for specialties other than their own. In general, many of us get much of our information about science—accurate or not—from stories in popular media. Illustrations accompanying these stories have perhaps the greatest impact on readers or viewers. Consider the number of times you have skimmed the headlines and looked at pictures and their captions before deciding whether or not to read a story.

In this article, we'll look at a few highlights from the extensive archives of dinosaur illustration, both scientific and popular, to see how well popular reconstructions have reflected scientific understanding of the appearance and behavior of these intriguing animals. We hope you will learn a bit about the interaction of dinosaur science and art—how popular representations have come to shape many of our changing perceptions of dinosaurs and their world.

Dinosaurs in Science and Art: Early Images

Paleontologists—particularly those who study dinosaurs—are luckier than scientists from some other disciplines. They tend to find a ready audience, eager to hear about the newest discoveries. People love to read about the biggest, the fastest, the toothiest, the strangest dinosaur yet found. Hence, information about dinosaurs and their relatives quickly finds space in newspapers and magazines. This seems to have been the case since dinosaurs were first recognized by science.

As early as 1676, accounts of large fossil bones show up in Western geological publications,[1] although illustrations are rare. These bones were usually attributed to gigantic human beings or other living animals, such as elephants, rather than extinct reptiles. The concept of "dinosaur" was unknown until the 1840s. William Buckland, an English minister and a professor of geology at Oxford University, wrote the first scientific description of what we now know as a dinosaur. He illustrated several *Megalosaurus* bones, including a piece of a lower jaw with teeth, and decided they had come from a large lizard.[2] *Megalosaurus* is now recognized to be a carnivorous dinosaur, similar to the more familiar *Tyrannosaurus*.

At about the same time (1825), Dr. Gideon Mantell, a physician in southern England, made sketches of the teeth of a "giant lizard" he named *Iguanodon*.[3] Mantell

Fig. 2.

Artistic rendition of a "modern" dinosaur, by Platt.

was also the first person to draw a picture of what such an animal might have looked like. He never published this sketch, but Fig. 3 is a reproduction of his picture.[4, 5]

Some of the most popular dinosaurs, including *Stegosaurus*, *Triceratops*, and "*Brontosaurus*," had their bones first scientifically illustrated in the latter part of the 19th century in papers by O.C. Marsh or E.D. Cope. Perhaps the most famous of these pictures is the sketch of a fairly complete "*Brontosaurus*" skeleton published by Marsh in 1883 (Fig. 4).[6] One problem: the skeleton had no head when it was discovered. Marsh wanted to show a complete skeletal reconstruction, so he based the skull in this sketch on other fossils found in the same area. He decided to use the head of another recently discovered sauropod dinosaur, *Camarasaurus*, as a model. *Camarasaurus* has a rather boxy looking skull with a blunt snout, which is what you see on Marsh's "*Brontosaurus*." Unfortunately, Marsh didn't tell his readers that this is what he had done, so nobody knew that the head on the "*Brontosaurus*" skeleton was pure conjecture. Such was Marsh's stature as a dinosaur expert that most (though not all) people accepted his reconstruction pretty much without question. When you look at pictures or models of "*Brontosaurus*" made between the 1880s and the 1970s, you will see the influence of Marsh's reconstruction. Almost all these brontosaurs are shown with short, square snouts. It wasn't until the 1970s, when

excavations in Dinosaur National Monument uncovered *Apatosaurus* skeletons with skulls, that we recognized that Marsh was wrong. "*Brontosaurus*" (now known as *Apatosaurus*) ac-

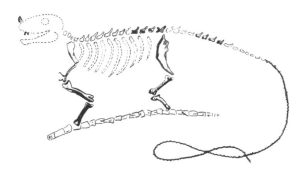

tually had a tapered head with a slender snout, much like its "cousin," *Diplodocus*.

Dinosaurs Go Public

So far, all of these early illustrations of dinosaurs come from the scientific literature. Who, then, gave us the first images of dinosaurs created primarily for a public audience? British painter John Martin may deserve that honor. Martin was an enormously popular 19th century painter, well known for his apocalyptic visions of biblical stories.[7] In the 1830s, his fame led to invitations to create frontispieces of the newly discovered "giant lizards" for popular

Fig. 3. Sketch by Platt after Mantell's sketch of *Iguanodon*.[4, 5] Note the "horn" on the nose. Later discoveries showed us that this feature is really a spike that belongs on the hand of the animal.

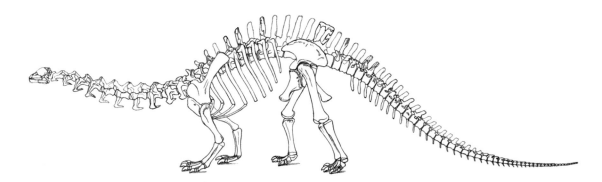

Fig. 4. Reconstruction of "*Brontosaurus*" (=*Apatosaurus*) by O.C. Marsh in 1883.[6] Bones illustrated with darker lines were drawn from the actual fossil material. Bones more lightly sketched are conjectures based on Marsh's knowledge of vertebrate anatomy. The head on this skeleton is incorrect (see text), as it belongs to another kind of sauropod, a *Camarosaurus*. This mistake was not corrected until the 1970s. Reprinted with permission of the Linda Hall Library, Kansas City, Missouri.

Fig. 5. An 1838 painting by John Martin, entitled *The Country of the Iguanodon*. In it, a *Megalosaurus* attacks an *Iguanodon*. Note the "horn" on the nose of the *Iguanodon*. The position of this spike is based on Mantell's original reconstructions. We now know the "horn" is actually a thumb spike, and belongs on the hand. A copy of this picture appeared as a frontispiece in *The Wonders of Geology*.[8] Reprinted with permission of the Linda Hall Library, Kansas City, Missouri.

geological publications by Mantell and others (Fig. 5).[8] Martin carried his flare for the melodramatic into these pictures of prehistoric landscapes. His dinosaurs, often shown frozen forever in mortal combat, writhe snake-like through exotic landscapes bedecked with strange-looking vegetation. Their resemblance to modern interpretations of dinosaurs is sometimes difficult to see! But, there is a sense of energy, motion, and vitality to these images that is often lacking from dinosaur illustrations later in the 19th and into the 20th centuries. In this respect, Martin's reptiles unknowingly foreshadowed the dynamic images of dinosaurs prevalent today.

British anatomist Sir Richard Owen (who coined the term "dinosauria")[9] and Benjamin Waterhouse Hawkins, an artist and sculptor, created what are arguably the most famous early representations of dinosaurs designed for public viewing. Hawkins was commissioned in 1853 to produce a diorama of prehistoric life for the grounds of the reconstructed Crystal Palace Exhibit in Sydenham, England (outside London)—sort of the first "Jurassic Park."

Owen acted as consultant for the project, providing scientific expertise on how the dinosaurs, including *Megalosaurus* and *Iguanodon*, might have appeared in life.[10] In reality, Owen had very few bones on which to base the appearance of the dinosaurs Hawkins was to build. Look carefully at Fig. 6. *Megalosaurus* doesn't look at all like modern reconstructions of large carnivorous dinosaurs, such as *Tyrannosaurus* or *Allosaurus*. First of all, it is standing on four legs, rather than two. In many ways, the body is more mammalian in appearance than it is reptilian. In particular, note how the legs do not sprawl out to the sides of the body, as they do in most reptiles. This sketch demonstrates how scientists may have to make interpretations based on limited physical data—though Owen also used comparisons with the anatomy of living organisms when making his interpretations, as do modern paleontologists.

The Crystal Palace reconstructions are also good examples of how personal and societal beliefs may influence scientific interpretation. Owen did not accept contemporary theories of evolution, particularly as exemplified by the concept of progressionism. This concept maintained that life on Earth evolved in a series of creation episodes, each preceded by a catastrophe that wiped out most of the organisms of the previous time interval. Animals and plants in each new episode were supposedly an improvement over those of the previous one. Progressionists viewed mammals, for example, as farther along the evolutionary road—more "advanced"—than reptiles.

Owen, one of the foremost anatomists of his day, had studied the body plans of numerous living vertebrates and correctly saw that dinosaurs were not simply scaled-up lizards, but quite a different sort of reptile. He wanted Hawkins' reconstructions of dinosaurs to show that dinosaurs were "more advanced," i.e., "more mammal-like" than modern reptiles. That way, he could use the fact that dinosaurs had gone extinct to "prove" that progressionist evolution ideas were wrong, as progressionism predicted that more advanced animals would not precede less advanced animals, nor could animals "de-evolve" to more primitive forms.

Fig. 6.

Hawkins' sketch of *Megalosaurus*,[10] onto which Owen drew the fossil bones that provided the basis for this reconstruction. Reprinted with permission of the Linda Hall Library, Kansas City, Missouri.

Although Owen's (hence Hawkins') dinosaurs were in some ways influenced as much by philosophical agendas as by science,[11, 12] Owen truly believed that his reconstructions were the most accurate possible, given the data available. And, it is worth noting that at least part of his reconstruction of dinosaur anatomy is now accepted. Modern illustrations of dinosaurs generally portray most of them with an upright, mammal-like stance, much as Owen did.

The point is that good scientists—and Owen really was a fine scientist—may make mistakes. Sometimes strongly held ideas are difficult to surrender; sometimes there is simply a lack of adequate data. Good science, however, is self-correcting. Eventually, with more study, better data, and new perspectives, we are able to revise our interpretations. We see such revisions when we look at dinosaur illustrations through time—the influence of paleontology's changing views of dinosaur anatomy and behavior.

Dimwits in Swamps: Other Early Images

In the late 1800s, the public learned that dinosaurs had also roamed North America. Remains of *Hadrosaurus* were discovered in New Jersey in 1858. These fossils gave science the first clues that not all dinosaurs lumbered about on four legs, as portrayed by Owen and Hawkins. Dr. Joseph Leidy of the Philadelphia Academy of Natural Science described *Hadrosaurus*, an ornithopod dinosaur similar to *Iguanodon*, as a biped, walking about on two legs, although he never illustrated this interpretation.[13]

In 1878, discovery of a collection of more than 30 *Iguanodon* skeletons, many of them nearly complete, added to the data suggesting to paleontologists that other ornithopod dinosaurs walked on two legs. Louis Dollo, a Belgian paleontologist, reconstructed a number of these skeletons, correcting in the process the "nose horn" mistake made by Mantell (Fig. 3). Still on display at the Royal Institute of Natural Science in Brussels, these mounted skeletons influenced illustrations of Iguanodons and their relatives for most of the next century.[14, 15, 16, 17]

In the late 1800s, Cope and Marsh were generating a new wave of dinomania with their discoveries in the American West. It is also around this time that the popular images of the gigantic "*Brontosaurus*," standing chest-deep in marshes, gave the Mesozoic world that distinctly "swampy" look that would characterize illustrations of The Age of Reptiles for years to come.[18, 19]

There seem to be relatively few surviving popular dinosaur illustrations from the very last part of the 19th century and early 20th century. Two notable

continued from p. 163

the President (2001–2002) of the Association for Women Geoscientists.

Brian Platt received his B.S. degree in Geology from Bloomsburg University of Pennsylvania. He is currently studying for his Masters in vertebrate paleontology at the University of Kansas. Besides his continuing work on dinosaurs in popular culture and paleoart, he plans to pursue his interests in taphonomy and paleoenvironmental reconstruction.

exceptions: Mary Mason Mitchell's sprawling, belly-dragging *Diplodocus*, and the famous works of Charles Knight.

Mitchell's drawing was commissioned by American paleontologist Oliver Hay for a paper he was writing.[19] Several prominent members of the paleontological community, O.C. Marsh, E.D. Cope, and H.F. Osborn among them, were beginning to view at least some dinosaurs as rather agile animals. Hay did not subscribe to this interpretation. He felt that an upright, mammal-like stance would be impossible, given the size of the abdominal regions of dinosaurs such as *Diplodocus*. Hence, he directed Mason to draw *Diplodocus* as a squat, serpentine beast, rather than a more elephant-like animal.[19, 20] This image is a bit of a throwback, reminiscent of illustrations of the "great lizards" of the early 1800s. Hay's interpretation was never accepted, but the image survives as an example of the controversy surrounding dinosaur reconstructions in the first part of the 20th century.

Charles Knight: Image-builder

Charles Knight may be the artist most responsible for creating the image of dinosaurs that many of us grew up with. He was a meticulous artist and sculptor, working closely with experts like Henry Fairfield Osborn to create reconstructions of *T. rex*, *Triceratops*, *Allosaurus*, and other familiar dinosaurs, giving generations of school children the now-classic images of the dinosaurs they love.

One of Knight's most famous paintings, a mural of *Triceratops* and *T. rex*, hangs in the Field Museum in Chicago. Although portraying the "ultimate" dinosaur confrontation, the scene is really pretty static. Knight had only scrappy material from two partial skeletons from which to work, but this interpretation of *T. rex* influenced artists for decades. Even today, buy a bag of those cheap little plastic dinosaurs; you'll likely find the *T. rex* looking more like those painted by Knight than the *T. rex* created by Spielberg animators!

Knight's paintings and sculptures solidified the concept of dinosaurs that held for most of the last century: sluggish behemoths, plodding

dim-wittedly toward their date with extinction. Stolid, lumbering dinosaurs are only part of Knight's story, however. His early works are often very exciting and full of motion. For example, Knight deserves credit for one of the

first truly modern portraits of dynamic dinosaurs—the battling theropods (*Dryptosaurus*) he painted in 1897 at the urging of E.D. Cope (Fig. 7). It would be 70 years before paleontology would again see dinosaurs as such nimble, agile animals. Although we can now see inaccuracies in Knight's portrayals, courtesy of the 20/20 vision of hindsight, his images represented the best of dinosaur science in his day. They certainly helped maintain public interest in the "fearfully great lizards" through the first half of the 20th century.

Perceptions Change

Public and scientific perceptions of dinosaurs, with few exceptions, changed little from the end of the 19th to the later part of the 20th centuries. People continued to view dinosaurs as

Fig. 7.

Charles Knight's very "modern-looking" reconstruction of two fighting *Dryosaurus* from 1898. Based on his 1897 painting of the same subject. Reprinted with permission of the Linda Hall Library, Kansas City, Missouri.

No. 3.   Megalosaurus (Lælaps, Dryptosaurus) aquiluuiguis (*Cope*).

big, dim-witted, rather sedentary creatures, doomed to fade into the last Cretaceous sunset. Rudolph Zallinger's famous mural at the Yale Peabody Museum, completed in 1947, beautifully reflects this view of dinosaurs. If you went to elementary school in the 1950s or 1960s, you have probably seen this painting—you may even have had a poster of it in your classroom.

Zallinger did paint some activity in his mural, but, overall, life in the Mesozoic seems rather quiet. You see the giant "*Brontosaurus*," standing placidly, hip-deep in a Jurassic swamp, ferny vegetation hanging from its mouth. *Allosaurus* feeds calmly on a bloody carcass. *Stegosaurus* stands nearby, apparently totally unperturbed by the proximity of the large carnivore. Overhead, pterosaurs hang frozen in the sky on silent, leathery wings. There is little sense of the exciting, dynamic dinosaurs of Knight's early work, or even of the frantic drama of Martin's snaky beasts of the 1830s.

Contrast the Zallinger portrayal of dinosaurs with the flocking *Gallimimus*, clever, claw-tapping *Velociraptors*, and lawyer-eating *T. rex* of *Jurassic Park*, or the artwork of Gregory Paul, Mark Hallett, or Karen Carr, among others. Quite a change! Everywhere you look today, illustrators, museum model makers, and computer animators portray active, complex behaviors in their dinosaurs. How did we get from "big and dumb" in the mid-1900s to "fast and smart" in the late 1900s? Blame it on *Deinonychus* (Fig. 8).

Blame It on *Deinonychus*

Deinonychus was discovered by Yale paleontologist John Ostrom in Montana in 1964.[21] If you like dinosaurs, you know *Deinonychus*. You've seen its Hollywood version portrayed as the *Velociraptors* in *Jurassic Park*. There may be no more evocative modern image of dinosaurs than Robert Bakker's portrait of *Deinonychus*, created for Ostrom's paper. Here is a dinosaur as an agile predator: tail straight out, grace and power evident in every line (Fig. 8). Ostrom wanted to illustrate

his hypothesis that some dinosaurs were far more bird-like than reptile-like in their behavior. Bakker's drawing clearly conveys this idea. Others in the past had hypothesized a bird-like behavior for dinosaurs,[22] and others had created images that hinted at a dynamic nature for these animals (Fig. 7).[23] But, it is the animal in Bakker's drawing, backed by Ostrom's detailed interpretation of the functional morphology (how anatomy relates to behavior) of *Deinonychus*, that finally overcame the years of "dinosaur as dumpy dim-wit." Thanks to *Deinonychus*, it is difficult today to find a picture of a dinosaur that isn't leaping, slashing, rearing, running, parenting, or munching on the odd bystander!

Beyond the Image

Today we have a greater understanding and appreciation of the complex anatomy, physiology, and behavior of dinosaurs. The discoveries made by scientists have been translated with at least a fair degree of accuracy into popular images. And those images are everywhere—books, magazines, museums, movies, television, the Internet. Yet, misconceptions about dinosaurs persist. Modern images may portray agile, dynamic animals, but when asked to describe dinosaurs, many people still use words like "big, large, not so smart, dumb, stupid, and cold-blooded," terms that reflect a 1950s paleontology. Somehow the science behind the images isn't always translating to the public. Maybe, when it comes to dinosaurs, we need to go beyond the pictures and read the whole story!

Fig. 8.

Deinonychus, by Platt, after the now-classic sketch by Bakker (1969).

References

1. Plot, R. 1676. *The natural history of Oxfordshire*. Oxford.

2. Buckland, W. 1824. Notice on the *Megalosaurus* or great fossil lizard of Stonesfield. *Transactions of the Geological Society of London*, series 2 1:390–396.

3. Mantell, G. 1825. Notice on the *Iguanodon*, a newly discovered fossil reptile, from the sandstone of Tilgate Forest, in Sussex. *Philosophical Transactions of the Royal Society of London* 115:179–186.

4. Lucas, S.G. 2000. *Dinosaurs: the textbook*, 3rd ed. McGraw-Hill.

5. Norman, D. 1988. *The prehistoric world of the dinosaur*. New York: Gallery Books.

6. Marsh, O.C. 1883. Principal characters of American Jurassic dinosaurs, Part VI: restoration of *Brontosaurus*. *American Journal of Science*, series 3 26:81–85.

7. Feaver, W. 1975. *The art of John Martin*. Oxford: Clarendon Press.

8. Mantell, G. 1838. *The wonders of geology*. London: Relfe and Fletcher.

9. Owen, R. 1842. Report on British fossil reptiles, Pt. II. *Reports of the British Association for the Advancement of Science* 11:60–204.

10. Owen, R. 1854. *Geology and inhabitants of the ancient world*. London: Crystal Palace Library, and Bradbury & Evans.

11. Torrens, H. 1997. Politics and paleontology: Richard Owen and the invention of dinosaurs. In *The complete dinosaur*, eds. J.O. Farlow and M.K. Brett-Surman, 175–190. Bloomington: Indiana University Press.

12. Gould, S.J. 1998. An awful, terrible dinosaurian irony (19th-century British anatomist Richard Owen and the etymology of dinosaurs). *Natural History* (Feb.).

13. Leidy, J. 1858. Remarks concerning *Hadrosaurus*. *Proceedings of the Academy of Natural Sciences of Philadelphia* 10:215–218.

14. Lydekker, R. 1896. *The royal natural history*. Volume V [Reptiles]. London: Warne.

15. Small, A.S. 1924. *The boys' book of Earth*. New York: E.P. Dutton & Co.

16. Swinton, W.E. 1969. *Digging for dinosaurs*. New York: Young Readers Press, Inc.

17. Steel, R., and A. Harvey. 1979. *The encyclopedia of prehistoric life*. New York: Gramercy Publishing Co.

18. Osborn, H.F. 1904. Fossil wonders of the West: the dinosaurs of the Bone-Cabin Quarry, being the first description of the greatest find of extinct animals ever made. *Century Magazine* 68:680–694.

19. Hay, O.P. 1910. *Proceedings of the Washington Academy of Sciences*, vol. 12, pp. 1-25.

20. Glut, D.F. 1980. *The dinosaur scrapbook*. Secaucus, NJ: The Citadel Press.

21. Ostrom, J. 1969. A new theropod dinosaur from the Lower Cretaceous of Montana. *Postilla* 128:1–17.

22. Huxley, T.H. 1868. On the animals which are most nearly intermediate between birds and reptiles. *Geological Magazine*, v. 5, pp. 357–365.

23. Knight, C.R. 1942. Parade of life through the ages. *National Geographic* 81(2):141–184.

Dinosaurs in the Movies: Fiction and Fact

20

Donald F. Glut

President, Frontline Entertainment, Inc.
Hollywood, CA

Donald F. Glut

is a freelance writer and motion-picture director who has had a passion for paleontology since he was seven years old. He has a B.A. degree in Letters, Arts and Sciences for Cinema from the University of Southern California. His writing credits include numerous books on dinosaurs, including the original *Dinosaur Dictionary* and the ongoing *Dinosaurs: The Encyclopedia* series of reference books. Perhaps he is best known for his novelization of the movie *The Empire Strikes Back*. Currently he is President of Frontline Entertainment, www.FrontlineFilms.com, and also a volunteer fossil preparator at the Natural History Museum of Los Angeles County.

A gigantic *Tyrannosaurus rex* stalks into a jungle clearing, the three-clawed fingers of each hand grasping at the air, its long tail swishing snake-like along the ground. The creature's eyes, set about midway along the head, are ever alert, seeking out prey and natural enemies. Pausing in its tracks, the dinosaur lowers its head to scratch its ear, then stomps, tail dragging behind it, toward its potential victim, a beautiful young woman cowering on the limb of a tree. This *Tyrannosaurus* remains unaware that it will soon be engaged in battle for its life against a 40-foot tall "prehistoric" gorilla.

In a different situation, yet another three-fingered *T. rex* stalks more formidable prey: a ponderously massive *Stegosaurus*, dragging behind it a long, heavy tail fortified with four deadly spikes. Watching this encounter is a variety of ancient reptiles—fin-backed pelycosaurs, horned ceratopsians, and long-necked sauropods—a silent saurian audience observing a prehistoric sporting event. Within moments the *Tyrannosaurus* and *Stegosaurus* will be, as the commonly used phrase puts it, "locked in mortal combat."

Scenes like these never actually occurred in "real" life, of course. On the other hand, they have occurred many times in "reel" life…in motion pictures.[1,2] Both scenes above are from classic films released by RKO Radio Pictures. The first, *King Kong* (1933), features three-dimensional models sculpted by Marcel Delgado. It was photographed one frame at a time by "stop motion" pioneer Willis O'Brien.[3] The other, *Fantasia* (1940), showcases the artistry of Walt Disney's team of animators.[4]

> *… many film-makers … seem to view [dinosaurs] as no more than fanciful monsters and, as such, no more bound to reality than are space aliens and supernatural creatures.*
>
> *Don Glut*

Misconception and Misrepresentation

While people like Delgado and O'Brien correctly regarded dinosaurs as real *animals* subject to natural laws, many film-makers, unfortunately, seem to view them as no more than fanciful *monsters* and, as such, no more bound to reality than are space aliens and supernatural creatures.[5] Some have treated

their dinosaur stars as little more than big, lizard-like monsters, existing only to lumber into a scene and fight it out with some other similar creature. This attitude has helped to perpetuate a number of misconceptions and myths about dinosaurs.

One particularly popular and prevalent misconception is that all extinct animals—from the most primitive reptiles to dinosaurs to Ice Age mammals—coexisted in some vague and undefined "prehistoric world."[6] Secondly, since the earliest days of the movies, dinosaurs have been incorrectly linked with "cavemen" in the latters' Stone Age world. Although no ancient human ever encountered a living dinosaur (the last nonavian dinosaurs and first humans being separated by a gap of more than 60 million years), their on-screen conflicts have provided much excitement and melodrama. Perhaps, mixing dinosaurs with early man is basically a matter of economics. Movies like *One Million B.C.* and its 1967 remake *One Million Years B.C.* (Fig. 1) were quite successful. Conversely, "Stone-Age" movies without dinosaurs, such as *Creatures the World Forgot* (1971), have usually done comparatively poorly at the box office.

Lend Me a Lizard

Another common misconception reflected in dinosaur movies involves posture and locomotion. Real dinosaurs walked with an upright

Fig. 2. A photographically enlarged rhinoceros iguana lizard menaces cave couple Tumak (Victor Mature) and Loana (Carole Landis) in the "Stone Age" melodrama *One Million B.C.* (1940).

stance rather than the sprawling posture displayed by living lizards and crocodilians. Nevertheless, live iguanas, tegus, monitors, chameleons, skinks, and even alligators have doubled as film dinosaurs. Sometimes these "performers" were outfitted with rubber horns and other embellishments to make them appear, at least in the imaginations of filmmakers, more "prehistoric."[7]

As early as 1913, an alligator was dressed up with various accoutrements to give it an archaic look for the D.W. Griffith film *Brute Force*. In the 1940s, most of the animal cast of *One Million B.C.* consisted of live lizards and alligators going through their sluggish, tail-dragging paces on miniature sets, and photographed in slow motion by special-effects man Fred Knoth (Fig. 2). Unfortunately, no matter how many horns and fins one sticks on these reptiles, lizards and alligators do *not* look like dinosaurs. Still, that fact has not deterred filmmakers from passing off their scaly pets as "dinosaurs," sometimes with astounding audacity and hilarious results. For example, the 1955 independent movie *King Dinosaur* offered a photographically enlarged green iguana

Fig. 1. *Ceratosaurus* combats *Triceratops* (models sculpted by Arthur Hayward and animated by Ray Harryhausen) as cave people flee in terror, all from different time periods and lumped together in a generalized "prehistoric world," in *One Million Years B.C.* (1967).

propped up into an awkward bipedal stance via an unseen wire. The first remake of *The Lost World* (1960) by 20ᵗʰ Century-Fox offered a similarly suspended monitor lizard, decked out with cranial horns and a *Dimetrodon*-like dorsal fin. In both cases, the uncomfortable reptile was identified on screen as a *Tyrannosaurus rex*, no doubt eliciting some interesting responses from theater audiences.

When a Lizard Just Won't Do, Get a Human!

Live reptiles, however, are not the best or most reliable species of actors. Consequently, ever since the silent days of motion pictures, dinosaurs have also been played on screen by human performers in dino-suits. Unfortunately, because the contours of a human being do not match those of any kind of dinosaur, movies featuring these animals have, on occasion, distorted dinosaur shapes to accommodate human actors stuffed inside the costumes. The results are not only scientifically ridiculous, but often extremely—if unintentionally—comical. Picture a very short, still three-fingered, *Tyrannosaurus* portrayed by stuntman Paul Stader in the movie *One Million B.C.* This *T. rex* attempts to menace the caveman hero, played by none other than Victor Mature, who was nearly at tall as the *T. rex!*

Human beings wearing stiflingly hot dinosaur suits (Fig. 3) also appeared in the 1948 movie *Unknown Island*. In this film, a group of horn-bearing, meat-eating dinosaurs were referred to as *Tyrannosaurus*, though they bore little resemblance to the real thing. They lumbered about awkwardly on screen. The costumes were made of latex-covered canvas, and under the blazing desert sun of Palmdale, California, they proved too much for at least one actor—the poor fellow dropped over of heat exhaustion.[8]

Early Dinosaur Stars
Gertie (1912)

Here's a great bit of trivia for your next party: the first character ever designed for an animated cartoon was a dinosaur. *Gertie*, the title character of the classic 1912 short film by cartoonist Winsor McCay, turns out to be a fairly "modern" depiction of a sauropod. Granted, Gertie—an *Apatosaurus*—had certain human characteristics, and lived in a world that included the woolly mammoth, but in other respects, McCay got much of his science correct. Gertie was portrayed as an active, herbivorous, fully terrestrial animal, rather than a sluggish, water-bound dimwit. She was also drawn by McCay with the correct, long, low head we now know this kind of dinosaur possessed, rather than the "boxlike" head in vogue for brontosaurs in the early 1900s.

© 1948 Film Classics

Fig. 3. Aping King Kong, a *Ceratosaurus* squares off with the giant ground sloth *Megatherium*, both creatures portrayed by human actors (the sloth by Ray "Crash" Corrigan) wearing uncomfortable costumes, in the motion picture *Unknown Island* (1948).

King Kong (1933)

The dinosaurs and other Mesozoic reptiles in the 1933 version of *King Kong* were also presented with some degree of anatomical accuracy, despite producer Merian C. Cooper's insistence to "Make [the beasts] bigger!" The prehistoric creatures in this film were the work of sculptor, Marcel Delgado, and photographer, Willis O'Brien, who also worked together in the classic silent movie *The Lost World* (1925). To ensure an accurate portrayal, Delgado and O'Brien used as reference the best and most up to date visual sources available to them—they studied the works of Charles R. Knight, a well-known and respected artist of prehistoric life.[9]

Knight, an amateur paleontologist, worked closely with the dinosaur specialists of his day, most notably Edward Drinker Cope, Henry Fairfield Osborn, and William Berryman Scott. Knight incorporated in his drawings, paintings, and sculpture the latest paleontological information, based upon fossil materials then available. He also drew on his own considerable knowledge of living animals—their skeletal structure, musculature, coloration, and behavior—to fill in aspects of dinosaur anatomy that were not yet known. His "educated guesses" were often later shown to be correct. Knight's works were, in their day, "state of the art" and scientifically accurate within the context of then—current knowledge. In that sense, the Mesozoic creatures in *King Kong* were quite correct.

Knight's works inspired the designs of most of the giant reptiles in *King Kong*. In the *Tyrannosaurus* versus Kong battle (Fig. 4), the dinosaur was based directly upon a small oil painting that Knight produced around 1906 for the American Museum. This painting—reproduced countless times since—depicts a three-fingered, tail-dragging creature confronting a family of *Triceratops*.[10] Unknown to Knight (or

later to Delgado and O'Brien when they began their work on *King Kong*), the early skeletal reconstructions of *Tyrannosaurus* upon which he based this painting were not entirely correct.

Knight's depiction of *T. rex* arose from his collaboration with paleontologist H.F. Osborn. Osborn first named and described *Tyrannosaurus* based on some fossil bones from Montana. Because this skeletal material was incomplete, Osborn modeled parts of *Tyrannosaurus* on the more completely known genus, *Allosaurus*, a related theropod. One result of this "borrowing of parts" was that Osborn gave his *T. rex* a three-fingered hand. We now know that the hands of the tyrannosaurs in *King Kong*, *Fantasia*, and other movies yet to come, sport one digit too many.[11, 12, 13, 14, 15] Another inaccuracy perpetrated by using *Allosaurus* as a model for *Tyrannosaurus* is the addition of about 10 feet of extra tail length to early reconstructions. *Tyrannosaurus* would later be portrayed as having a considerably shorter tail than that shown on the early skeletal mounts and in Knight's original painting.[16] We have also since learned that the tail of *Tyrannosaurus* was incapable of the smooth serpentine action animated by O'Brien.[17] Osborn also apparently misidentified the antorbital fenestra—a large opening about midway down the length of the skull—as the orbit or eye socket of *Tyrannosaurus*, resulting in an incorrectly placed eye in Knight's painting.

Osborn did have Knight pose his *Tyrannosaurus* in an upright, basically "manlike" position, rather than in a more lizard-like sprawl. It would not be until the 1960s and 1970s that dinosaurs would be described in popular texts as carrying their tails off the ground.

The other Mesozoic animals of *King Kong* also deviated somewhat from reality. The herbivorous *Apatosaurus*, for example, was portrayed as a swamp-dweller, as such gigantic sauropods were then thought to be. We now know that *Apatosaurus* and its relatives were fully terrestrial animals, despite their large sizes.

Other behaviors we now recognize as inaccurate are also portrayed in this film. For example, some of the dinosaurs in *King Kong* appear to be suffering from delusions about their normal diet—herbivorous *Apatosaurus*

Fig. 4. A giant "prehistoric gorilla" battles an oversized *Tyrannosaurus* (the latter based on a painting by Charles R. Knight), sculpted by Marcel Delgado and animated by Willis O'Brien for the motion picture *King Kong* (1933).

pursues fleeing crewmen, apparently with a dinner of human being in mind. And the marine reptile *Elasmosaurus* turns and twists its body mimicking an enormous snake. Such behaviors, however, were not the fault of Delgado and O'Brien, as they merely supplied what the script called for.

Fantasia (1940)

The creative forces behind the animated masterpiece *Fantasia* were concerned as much with drama and artistry as they were with scientific accuracy. Certainly, *Fantasia*'s version of the Mesozoic Era is a kind of "catch all" of time and geography. Many species of extinct reptiles—some of them separated by a hundred or more million years in reality, not to mention a major ocean basin or two–are all depicted as though they lived together. The *Tyrannosaurus* and *Stegosaurus* that confront one another on screen could never have met in battle, or for any other reason. More than 70 million years actually separate the death of the last *Stegosaurus* in the Late Jurassic from the hatching of the first *Tyrannosaurus* in the Late Cretaceous.

The *Tyrannosaurus* in *Fantasia* is, not surprisingly, inaccurately portrayed. It is drawn with forelimbs that are longer and more robust than they really were, possibly based upon an *Allosaurus* model. It is again depicted with three digits on each hand. This time, however, blame cannot be placed on inaccurate data or on Charles Knight. Reportedly, Walt Disney insisted that his *Tyrannosaurus* in *Fantasia* possess three fingers instead of the correct two, because three "looked better." (Disney company employees would continue to portray *Tyrannosaurus* with three-fingered hands in subsequent studio projects for decades after Disney's death; the full-sized "audio-animatronic" figure built for the 1964 New York World's Fair and the mechanical tyrannosaur appearing in the 1985 movie, *My Science Project*, are just two examples.)

Unknown Island (1948)

Charles Knight was also the inspiration behind other movie dinosaurs. *Unknown Island* was inspired by an article he authored and illustrated for the February 1942 issue of the *National Geographic* magazine. This "inspiration," however, did not translate very well to the silver screen, as the dinosaurs and other extinct animals in the movie—including the swamp-bound sauropod *Diplodocus* and crawling fin-backed pelycosaur *Dimetrodon*—were presented either as clumsy, mechanically operated "puppets" or, more often, as people in dino-suits. The horned theropod, *Ceratosaurus*, from Knight's article didn't exactly make it into the movie, although its horns did—on a group of lumbering meat-eater (as mentioned above) identified as "tyrant dinosaurs."

Fact *versus* Fantasy in Early Dino Movies

In this writer's opinion, two things are evident from the above examples:

1) Delgado and O'Brien did their best to ensure anatomical accuracy in their creations, based on the best scientific information then available to them. This effort is even more obvious in the *Lost World* (Fig. 5), where groups of the same dinosaur species are shown interacting (as opposed to the one-individual-per-genus format of *King Kong*). Arguably, *The Lost World* is almost a film tribute to Knight and his ideas about dinosaurian anatomy and behavior. His paintings and sculptures seemingly come alive through the work of Delgado and O'Brien.

2) Although the artistry in *Fantasia* is superb, the "science" presented in the film is replete with misinformation, even by 1940 standards. True, filmmakers are not scientists, and the majority of them may have little knowledge of, or even concern for, the facts. Yet, explaining the lack of accuracy in the dinosaur segment of *Fantasia* is a bit difficult—particularly considering that American Museum of Natural History paleontologist Barnum Brown was an official consultant for the film. The example of inaccuracies described above suggests that many of his suggestions were ignored.

"Modern" Movie Dinos
Jurassic Park (1993)

The dinosaur movie genre that was so popular in previous decades was virtually "reborn" in

the late 20th century, thanks to the development of a highly sophisticated special effects process called Computer Generated (or Graphics) Imagery, or CGI. Computer-based visuals allow filmmakers literally to show anything on the screen that they can imagine, "living" dinosaurs included. CGI dinosaurs, perfected by the techno-wizards at Industrial Light and Magic (ILM), debuted in the 1993 blockbuster *Jurassic Park* (Fig. 6), directed by Steven Spielberg and based on the Michael Crichton novel (1990) of the same title.[18]

With the premiere of *Jurassic Park*, movie dinosaurs were no longer the clunky, tail-dragging "monsters" of earlier movies. At last, their looks and behaviors were based mainly on modern scientific data. Dinosaurs were finally portrayed in the active, tail-off-the-ground posture/stance reflective of current paleontological thinking. Much of this dinosaurian "new attitude" can be traced to the "Dinosaur Renaissance"[19] of the mid-1970s and, in the particular case of *Jurassic Park*, to paleontologist John R. Horner of the Museum of the Rockies, who was brought on board as a dinosaur consultant.

Despite Horner's best efforts to ensure accuracy in this film, the director indulged in certain dramatic liberties with the dinosaurs. While the *Jurassic Park* dinosaurs walked with correct postures, they fell short in other respects. Regrettably, some of these alterations made on the dinosaurs in this film were readily accepted by viewers to be scientific fact. For instance, the small Asian theropod *Velociraptor* was physically patterned after its substantially bigger North American relative *Deinonychus*. Hence, it was depicted as an animal several times larger—and probably many times more intelligent—than its real-life counterpart. By contrast, the large theropod *Dilophosaurus* was shown as an animal considerably smaller than the real-life version, and one equipped with both a flaring, lizard-like neck frill and the ability to spit poison saliva. Neither of these "adaptations" is based on scientific data.

Jurassic Park also introduced the term "raptor" into the popular lexicon. This suffix, meaning "robber," is applied in the paleontological literature to various kinds of carnivo-

© First National

Fig. 5. Accurate (in their day) *Allosaurus* and *Triceratops* models based on the artwork of Charles R. Knight, sculpted by Marcel Delgado and animated by Willis O'Brien, for the movie *The Lost World* (1925).

rous dinosaurs, such as the very primitive saurischian *Eoraptor*, the allosauroid *Sinraptor*, the oviraptorid *Oviraptor*, and the dromaeosaurid *Bambiraptor*. Following the release of *Jurassic Park*, the term "raptor" began to be misused, by laymen and some scientists alike. It often describes not only members of the Dromaeosauridae (the theropod family to which both *Velociraptor* and *Deinonychus* belong), but just about any kind of theropod dinosaur bearing a sickle-like claw on its foot.

The success of *Jurassic Park* opened the proverbial door for other projects featuring more authentic-looking, more realistically behaved, computer-generated dinosaurs. There were, naturally, the sequels: *The Lost World: Jurassic Park* (1997) and *Jurassic Park III* (2001). The original movie, and the visual effects it pioneered also inspired numerous other films and television productions. *Walking with Dinosaurs*, *Allosaurus*, and *When Dinosaurs Roamed America* have all appeared on The Discovery Channel within the last few years. In some cases, these programs include speculation presented more or less as "fact," yet they do give audiences a true sense of dinosaurs as living, breathing animals. Part of this effect is due to the method of filming; the programs are shot

as if photographed by nature cameramen, shooting authentic footage of live animals in the wild.

Dinosaur (2000)

Perhaps the most ambitious and costly of all post-*Jurassic Park* projects to date has been the Disney studio's *Dinosaur* (2000), a computer-created motion picture epic that was in various stages of planning, pre-production, production and post-production for well over a decade. Visually stunning, and incorporating the convention that animals can talk, this movie was about a young *Iguanodon* and his efforts to lead his dinosaur friends to a valley where they can live in peace.

Unfortunately, this highly anticipated film continued to offer much misinformation about the way dinosaurs looked and behaved, even dredging up clichéd misconceptions first produced on film in the days of silent pictures. Once again animals from different geologic time periods and geographic locations were depicted coexisting in some generalized "prehistoric world." And, perhaps in a nod toward "political correctness," meat-eating dinosaurs, particularly those of the horned genus *Carnotaurus*, were portrayed as non-speaking, virtually mindless, outright villains, rather than normal animals that make their natural living by eating the flesh of other animals. In contrast, modern studies appear to indicate that theropods ranked relatively high on the scale of dinosaurian intelligence.

There are problems with the portrayal of the movie's herbivorous dinosaurs, as well. Some of the older *Iguanodon* individuals are depicted as massive, entirely quadrupedal, almost rhinoceros-like animals, possessing horn-like growths above their noses. In fact, they resemble quite strikingly the original life restorations of *Iguanodon* made by sculptor Benjamin Waterhouse Hawkins for the Crystal Palace exhibit back in 1853![20]

Art Eventually Imitates Life?

Sometimes it takes a long time for art to reflect reality. Much has been learned about dinosaurs—both in their physical appearance and possible behavior—since the first remains of these animals were collected and identified during the first quarter of the 19th century. Many of our newer interpretations have been based on discoveries made just in the past few decades.[21, 22, 23, 24, 25] Some very intriguing hypotheses have come out of this recent research.

© & ™ Universal Pictures and Amblin Entertainment

Fig. 6. Although this *Tyrannosaurus* head is really a full-sized mechanical mock-up created by special-effects master Stan Winston, most of the dinosaurs in the blockbuster movie *Jurassic Park* (1993) were generated by computers.

These include the ideas that at least some kinds of dinosaurs may have been warm-blooded, that the bodies of some theropod dinosaurs were covered with feathers or feather-like structures, that some dinosaurs enjoyed comparatively sophisticated social activity, and that at least one group of theropods evolved into birds. How different these concepts are over the older notions—certainly interesting and exciting in their own right—of the sluggish, tail-dragging, lizard-like "mon-

sters" depicted in older texts, illustrations, and pre-*Jurassic Park* motion pictures.

References

1. Shapiro, M. 1992. *When dinosaurs ruled the screen*. New York: Image Publishing.

2. Jones, S. 1993. *The illustrated dinosaur movie guide*. London: Titan Books.

3. Goldner, O., and G.E. Turner. 1975. *The making of King Kong*. Cranbury, NJ: A.S. Barnes and Co., Inc.

4. Culhane, J. 1983. *Walt Disney's Fantasia*. New York: Harry N. Abrams, Inc.

5. Glut, D.F. 1997. Dinosaurs: real vs. reel. *Dinosaur* issue, *Starlog* magazine.

6. Rudwick, M.J.S. 1992. *Scenes from deep time: early pictorial representations of the prehistoric world*. Chicago and London: The University of Chicago Press.

7. Glut, D.F. 1980. *The dinosaur scrapbook*. Secaucus, NJ: Citadel Press, a division of Lyle Stuart, Inc.

8. Burns, B., with J. Michlig. 2000. *It came from Bob's basement*. San Francisco: *Chronicle Books*.

9. Czerkas, S., and D.F. Glut. 1982. *Dinosaurs, mammoths and cavemen: the art of Charles R. Knight*. New York: E.P. Dutton, Inc.

10. Colbert, E.H. 1945. *The dinosaur book: the ruling reptiles and their relatives*. New York: American Museum of Natural History, Man and Nature Publications Handbook, no. 14.

11. Lambe, L.M. 1914. On the forelimb of a carnivorous dinosaur from the Belly River Formation of Alberta, and a new genus of Ceratopsia from the same horizon, with remarks on the integument of some Cretaceous herbivorous dinosaurs. *Ottawa Naturalist* (27) 10:129–135.

12. Osborn, H.F. 1917. Skeletal adaptations in *Ornitholestes*, *Struthiomimus* and *Tyrannosaurus*. *Bulletin of the American Museum of Natural History* 35:733–771.

13. Carpenter, K., and M.B. Smith. 1995. Osteology and functional morphology of the forelimb in tyrannosaurids as compared with other theropods (Dinosauria). *Journal of Vertebrate Paleontology*, Abstracts of Papers, Fifty-sixth Annual Meeting.

14. Gilmore, C.W. 1920. Osteology of the carnivorous dinosauria in the United States National Museum, with special reference to the genera *Antrodemus* (*Allosaurus*) and *Ceratosaurus*. *Bulletin of the United States National Museum* 110:1–154.

15. Madsen, J.H., Jr. 1976. *Allosaurus fragilis*: a revised osteology. *Utah Geological Survey Publications*.

16. Newman, B.H. 1970. Stance and gait in the flesh-eating dinosaur *Tyrannosaurus*. *Biological Journal of the Linnean Society* (2) 2:119–123.

17. de Camp, L.S., and C.C. de Camp. 1968. *The day of the dinosaur*. New York: Doubleday and Company, Inc.

18. Crichton, M. 1990. *Jurassic park*. New York: Alfred A. Knopf.

19. Bakker, R.T. 1975. Dinosaur renaissance. *Scientific American* (232) 4:58–78.

20. McCarthy, S., and M. Gilbert. 1994. *The Crystal Palace dinosaurs: the story of the world's first prehistoric sculptures*. London: The Crystal Palace Foundation.

21. Desmond, A.J. 1976. *The hot-blooded dinosaurs: a revolution in palaeontology*. New York: The Dial Press/James Wade.

22. Bakker, R. 1986. *The dinosaur heresies: new theories unlocking the mystery of the dinosaurs and their extinction*. New York: William Morrow and Company, Inc.

23. Weishampel, D.B., P. Dodson, and H. Osmólska. 1990. *The dinosauria*. Berkeley: University of California Press.

24. Currie, P.J., and K. Padian. 1997. *Encyclopedia of dinosaurs*. San Diego: Academic Press.

25. Farlow, J.O., and M.K. Brett-Surman. 1997. *The complete dinosaur*. Bloomington: Indiana University Press.

www dinosaurs

The Internet is an excellent source for additional information about dinosaurs and paleontology in general. However, if you use a search engine using the word "dinosaurs" you will be inundated with possible choices. How do you choose? How can you tell if the sites are scientifically accurate and current? As with any web search, this is difficult. The following is a partial listing of the favorite sites of several members of the Society of Vertebrate Paleontology, listed alphabetically. Each site may well link you to others, which you may find interesting. **Have fun!**

Tracy L. Ford and Judith G. Scotchmoor

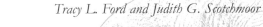

◆ ◆ ◆

Dinobase presents a dinosaur database including a list of dinosaurs, a classification of dinosaurs, pictures, and more…. The pages are created by Samira Cuny and Mike Benton of the University of Bristol.

> http://palaeo.gly.bris.ac.uk/dinobase/dinopage.html

Dinorama is a website of the National Geographic Society that includes information on what is new and exciting in dinosaur discoveries. Dinorama is a great place to enjoy vivid photographs and colorful artwork on topics such as dinosaur-bird evolution, the development of feathers, and dinosaur movement.

> http://www.nationalgeographic.com/dinorama/frame.html

Dinosauria on the University of California Museum of Paleontology website covers information on the fossil record, dinosaur life history and ecology, systematics and morphology. The site includes sections on dispelling dinosaur myths, recent dinosaur research and discoveries, and a listing of dinolinks.

> http://www.ucmp.berkeley.edu/diapsids/dinosaur.html

The Dinosauricon is a wonderful reference site. It maintains pages on every known dinosaur and their spatial distribution, includes several cladograms, and provides an art gallery with more than 65 artists.

> http://dinosauricon.com/

Dino Russ' Lair includes a host of detailed and updated information on dinosaur digs, dinosaur art, and dinosaur websites. The site is maintained by geologist Russ Jacobson of the Illinois State Geological Survey.

> http://www.isgs.uiuc.edu/dinos/dinos_home.html

Jurassic Park Institute provides updated news on dinosaur research and discoveries, dinosaur facts, and links to other resources, as well as an interactive DinoLab for student exploration.

http://www.jpinstitute.com/index.jsp

Project Exploration is a nonprofit education organization founded by Paul Sereno and Gabe Lyon. Within its website, the Expeditions section is a great place to explore and learn about the challenges involved in dinosaur expeditions in remote places, such as Inner Mongloia, Niger, Morocco, and Argentina. When it comes to getting the feel of being on a real dinosaur expedition, this site does a great job.

http://www.projectexploration.org/expeditions.htm

The Smithsonian Dinosaur Exhibits provides you with a glimpse of some of the dinosaur specimens on view at the National Museum of Natural History.

http://www.nmnh.si.edu/paleo/dino/

Sue at the Field Museum is an interesting virtual exhibit about the largest, most complete *T. rex* found thus far. The site provides an opportunity for you to get to know all about Sue, from her vital statistics to her discovery, significance, and even her controversy.

http://www.fmnh.org/sue/

Zoom Dinosaurs is created by Enchanted Learning, an organization that produces educational websites and games for children. As such, the site is easy to understand and includes information on dinosaur anatomy and behavior, a current news section, as well as features such as dinosaur quizzes and a dinosaur dictionary.

http://www.enchantedlearning.com/subjects/dinosaurs/

DinoHunter Dino Links provides a long list of dinosaur and paleontology sites far beyond those listed here.

http://www.dinohunter.info/html/links.htm

And beyond dinosaurs...

Follow a Fossil traces a fossil (vertebrate, invertebrate, or plant) from its discovery through its excavation, preparation, curation, research, and exhibition. This site is part of the Denver Basin Project at the Denver Museum of Nature and Science.

http://www.dmns.org/denverbasin2/fossil/index.html

The Tree of Life provides information about the diversity of organisms on Earth, their history, and characteristics. It may take a bit of exploration, but the effort is well worth it.

http://tolweb.org/tree/phylogeny.html

The **University of California Museum of Paleontology** maintains an extensive website providing a wide array of information and resources on phylogeny, geology, and evolution, as well as educational resources.

http://www.ucmp.berkeley.edu/

Glossary

analogous Having similar form and function, as in the wing of a bat and the wing of an insect, but not the same origin; Cf: *homologous*.

anticline An arched fold of layered (stratified) rock from whose central axis the strata slope downward in opposite directions.

assemblage Any group of organisms, fossil or living, found in the same place.

biogeography The study of the distributions of living organisms.

biomechanical The functions of an organism explained in the same terminology as a machine.

biota A collective term for all of the organisms in a given interval of time or in a given geographic area.

biotic Pertaining to life.

bipedal Walking on two feet.

canid A group of carnivores defined by features of the ear, commonly referred to as 'dogs'.

ceratopsians Any of the Suborder of Ceratopsia of horned ornithischian dinosaurs of the Cretaceous Period with a beaklike snout and a bony crest at the base of the head.

cetacean Any of the Order Cetacea; nearly hairless, fishlike water mammals, lacking external hind limbs, but having paddle-like forelimbs; includes whales, porpoises, and dolphins.

character Any distinctive feature or trait of an organism that can be used to infer its evolutionary history.

clade A higher taxon, e.g., a genus, family, phylum, that contains the ancestral species of the group and all its descendants.

cladistics A system of classification that uses the distribution of shared derived characters to determine ancestor-descendent relationships..

cladogram A branching diagram that illustrate patterns of evolutionary relationship; points of branching represent new features or characters that are shared by all of the taxa above the branch point. Cf: *phylogenetic tree*.

coprolites The fossilized excrement of animals.

correlation The process of determining age equivalence in rocks; fossils and radioisotopes are common means of doing correlations.

cross-bedded Having layers of rock oblique or transverse to the main beds of stratified rock.

derived character A feature or attribute possessed by a descendant taxon; this attribute has been modified from a homologous ancestral primitive character.

detritus Fragments of rocks produced during weathering and erosion by disintegration, abrasion, and other processes; essentially any debris.

Dimetrodon The sailed-backed pelycosaur of the Permian Period measuring up to 11.5 feet with enormously elongated neural spines used as temperature regulators; common in the red beds of Texas and Oklahoma.

dinosaur Any of the various extinct, Mesozoic Era, land-dwelling reptiles of the Saurischia or Ornithischia . In the "classical" sense (i.e., excluding their descendants, the birds), one of the most successful terrestrial groups to ever have inhabited our planet.

Diplodocus A genus of huge, long-necked, long-tailed, plant-eating sauropod dinosaur of the Jurassic Period.

diversity A measure or description of the numbers of kinds of organisms present in a given place or time.

dromaeosaurid Any of the agile, swift-moving, razor-toothed carnivorous theropods with a large brain, huge eyes, and a sickle-shaped talon on the inner toes, such as *Velociraptor* and *Deinonychus*.

ectotherm Refers to animals that use external energy sources to regulate internal body temperatures, often referred to as "cold-blooded."

endotherm Refers to animals that regulate their internal body temperatures without an external energy source, often referred to as "warm-blooded."

epicontinental Refers to a shallow sea that overlies a portion of a continental landmass; Hudson Bay is a modern example.

facultative bipedalism Capable of walking on hind limbs, but with the ability also to move on four limbs.

falsifiability The ability to formulate an hypothesis or statement in such a manner that empirical testing is possible, and such that the hypothesis or statement can be shown to be untrue.

flood basalts Dark, fine-grained, extrusive igneous rocks that are often found in vast sheets, as from a flow of lava.

fluvial Pertaining to or produced by a river.

fossil Any naturally occurring evidence of past prehistoric life.

gigantothermy The unique physiology associated with huge animals, characterized by a slightly elevated metabolic rate combined with large body size, the use of peripheral tissues as insulation, and active control of blood flow to retain or disperse heat from the body as appropriate.

Gondwana The land mass present during the Paleozoic Era that is now represented by South America, Africa, Australia, and Antarctica.

hadrosaurs Any of the Family Hadrosauridae, the large duck-billed ornithopod dinosaurs of the Cretaceous Period.

Haversian canal A channel through a bone that contains a blood vessel.

homologous Similarity of form (although not necessarily of function) between parts of different organisms; the result of evolutionary change from the corresponding part of a common ancestor. Cf: *analogous*. Homologous characters share common ancestry.

hypothesis An idea that is tentatively assumed, then scientifically tested for validity; in the vernacular, an "educated guess" based on past experience and/or knowledge.

ichthyosaurs Any of the extinct Order Ichthyosauria, prehistoric marine reptiles that had a fishlike body, four paddle-shaped flippers, and a dolphin-like head.

iridium A rare element that occurs in low levels in nature; higher concentrations of the element in rocks is usually attributed to increased volcanic activity or to extraterrestrial origins such as meteorites.

Laurasia The land mass present during the Paleozoic Era that is now represented by North America and Eurasia.

lineage [evol] A line of descent; a chronological succession of ancestor-descendant taxa.

lithosphere The outermost 100 kilometers of the Earth, composed of the crust and a portion of the upper mantle; the rigid outer shell of the Earth.

maniraptorans Small theropods with distinctive wrist design.

mass extinction An extinction event that eliminates a significant percentage of the existing organisms on Earth at the time.

Mesozoic The middle geologic era of the Phanerozoic Eon, subdivided into the Triassic, Jurassic, and Cretaceous periods, and characterized by the development and extinction of the dinosaurs, as well as the development of the first birds, mammals, and flowering plants (approximately 144 to 208 million years ago).

microstructure The microscopic structure of something; term often applied to the study of body tissues such as enamel and bone.

morphology The observable physical form or shape of living or fossil organisms or the study thereof.

mosasaurs A group of aquatic reptiles that appeared at the end of the Cretaceous Period and did not survive beyond that time, characterized by a streamlined body, short neck, long tail, and four paddle-like limbs.

mya Abbreviation for 'millions of years ago.'

Mysticete The baleen whales that appeared in the Oligocene and grow to enormous sizes, such as the blue whale, as opposed to the Odontocete or toothed whales.

ornithopod Any of the Suborder Ornithopoda of ornithischian dinosaur that walked upright on digitigrade (=walking on toes) feet.

paleoecosystem An ancient ecosystem, defined by the organisms and their interactions with one another and the environment.

paleoenvironment An ancient environment, defined by physical factors.

Pangaea A giant landmass formed by the collision of Gondwana with Laurasia during the Late Paleozoic.

paradigm An established model or theme.

paradigm shift Change in an established model or theme within a science that is the result of a significant scientific breakthrough.

phylogenetics The study of the evolutionary relationships of taxa through time.

phylogenetic tree A branching diagram that illustrates evolutionary relationships among taxa through time; points of branching represent ancestral populations.

phylogeny The line(s) of descent of a given taxon or taxa or the study thereof.

plate tectonics Theory of global-scale movements of the Earth's *lithosphere*; the lithosphere is divided into a number of large slabs (plates) whose horizontal movements produce seismic activity, i.e., earthquakes, and/or volcanism along their boundaries with other plates.

plesiosaurs Any of the extinct Order Sauropterygia, large water reptiles of the Mesozoic Era, characterized by a small head, long neck, short tail, and four paddle-like limbs.

pneumatic Containing air or air cavities.

primitive character In *cladistics*, an attribute or feature possessed by an ancestral taxon.

radiation [biol] The relatively rapid diversification of taxa following a move into a new environment or a major extinction event.

radioactivity Giving off energy in the form of particles or rays — such as alpha, beta, and gamma rays—by the spontaneous disintegration of atomic nuclei.

regression The lowering of sea level and the movement of the shoreline towards the ocean basin.

sauropod Any of the Superfamily Sauropoda, a type of gigantic, plant-eating, four-footed saurischian dinosaur with a long neck and tail, and a small head.

sedimentary Rocks formed by the deposition of particles of rocks or minerals.

settling velocity The speed achieved by an object — a sediment particle or bone, for example — dropping through a column of water.

shocked lamellae Submicroscopic parallel to sub-parallel deformation features in mineral crystals.

skeletal element A bone in the skeleton of an animal.

stratum A single layer of sedimentary rock. Pl: *strata*.

stratigraphy The study of layered rocks and sediments, strata; more generally, the interpretation of form, original age relationships, distribution, composition, and other properties of rocks as layers.

superposition The order in which rocks are placed or accumulated in beds, one above the other; in an undisturbed set of layers; the highest (uppermost) bed will be the youngest.

systematics The study of the evolutionary relationships among living and extinct organisms.

taphonomy A branch of paleontology: Specifically, the study of the processes that occur from the death of an organism through its discovery as fossil remains.

taxon A named group of organisms of any rank within a classification scheme used in biology and paleontology, for example, a species, an order, a sub-class, etc. may be the taxon under consideration. Pl: *taxa*.

taxonomy The theory and practice of classifying organisms. Sometimes used synonymously with *systematics*, however, the latter term has a broader connotation.

tetrapod Any vertebrate having four legs or limbs, including mammals, birds and reptiles.

thecodont Any taxon of the Order Thecodontia, reptiles of the Permian and Triassic periods believed to be ancestors of the dinosaurs and crocodilians.

theory A well-tested and substantiated explanation for some aspect of the natural universe; framework concept within which new hypotheses are formulated and tested, and against which newly acquired data are evaluated.

thermoregulation The control of body temperature at a constant level by processes of heat production, heat transport, etc.

trace fossil A preserved mark, such as footprint, burrow, or trail, left by the passage or activity of an animal.

varanid lizard Large lizards of southeastern Asia, such as the 13-foot-long Komodo dragon.

Western Interior Seaway An ancient epicontinental seaway of the interior of the United States during the Mesozoic Era.

woolly mammoth The group of extinct elephants with hairy skin and long tusks curving upwards. Remains have been found in North America, Europe, and Asia.

Index

Credits

Inside front cover — clockwise from top: Magnolia blossoms (stock photography), Ginkgo fossil (UC Museum of Paleontology), Early Jurassic mammal skeleton (modified from Jenkins and Parrington, 1976), Dragonfly fossil (Hemera), *Allosaurus* skull (Smithsonian Institution), *Archaeopteryx* (*Fossils, The Key to the Past*, R. Fortey), *Tyrannosaurus* and *Supersaursus* (Fabulous Fossils poster USDA Forest Service), Cycad fossil (D. Springer), Ammonite (G. James), Shark (stock photography), Paleoglobe — PALEOMAP Project (C. Scotese, University of Texas at Arlington) www.scotese.com.

Page iv — Fossil of *Sinosauropteryx* showing a covering of downy feathers (R. Walters and T. Kissinger), Sauropod trackways in southeastern Colorado (USDA Forest Service).

Page 2 — clockwise from top left: CT scanning of *Triceratops* limb bones (R. Chapman), Ichthyosaur restoration (N. Fraser), *Brachiosaurus* (clip art), Field team with humerus of *Paralititan stromeri* (P. Dodson), *Allosaurus* skeleton (D. Chure), Brushing fossil bones (M. Lindner).

Inside back cover — clockwise from top: *Pteranodon* (clip art), Morrison Formation, Wyoming, outcrop (D. Smith), *Triceratops sculpture*, 1/6 scale (R. Chapman), *Triceratops* skull (R. Chapman), *Sinosauropteryx* (R. Walters and T. Kissinger), Bone bed (D. Smith), Ammonite (G. James), CT image of juvenile hadrosaur (R. Chapman), CT- scanning of *Triceratops* limb bones (R. Chapman).